CONFESSIONS

OF A

ROGUE

NUCLEAR

REGULATOR

GREGORY B. JACZKO

Simon & Schuster

NEW YORK   LONDON   TORONTO   SYDNEY   NEW DELHI

Simon & Schuster
1230 Avenue of the Americas
New York, NY 10020

First Simon & Schuster hardcover edition January 2019

SIMON & SCHUSTER and colophon are registered trademarks
of Simon & Schuster, Inc.

For information about special discounts for bulk purchases,
please contact Simon & Schuster Special Sales at 1-866-506-1949
or business@simonandschuster.com.

The Simon & Schuster Speakers Bureau can bring authors to your
live event. For more information, or to book an event, contact
the Simon & Schuster Speakers Bureau at 1-866-248-3049
or visit our website at www.simonspeakers.com.

Interior design by Paul Dippolito

Manufactured in the United States of America

1   3   5   7   9   10   8   6   4   2

Library of Congress Cataloging-in-Publication Data
Names: Jaczko, Gregory B., 1970–   author.
Title: Confessions of a rogue nuclear regulator / Gregory B. Jaczko.
Description: New York : Simon & Schuster, [2019] | Includes index.
Identifiers: LCCN 2018013694 | ISBN 9781476755762 |
ISBN 9781476755786 (ebook)
Subjects: LCSH: Nuclear energy—United States.
Classification: LCC HD9698.U5 J33 2019 | DDC 333.792/40973—
dc23 LC record available at https://lccn.loc.gov/2018013694

ISBN 978-1-4767-5576-2
ISBN 978-1-4767-5578-6 (ebook)

CONTENTS

PROLOGUE

I never planned to be in a position to tell this story. A trained physicist, a Birkenstock-wearing PhD still amazed that a few simple equations could explain something as extraordinary as the northern lights, I never intended to become a nuclear regulator.

Before I came to Washington, I had never heard of the Nuclear Regulatory Commission. There are no television shows or movies with dashing federal agents rushing into a nuclear power plant with blue blazers flashing NRC logos. But because of a powerful politician and a right-place-at-the-right-time kind of timing, I became not only a nuclear regulator but the head of the agency.

This is how my first conversation with Harry Reid, the second most powerful Democrat in the Senate, who eventually got me on the commission, went back in 2001 when I was interviewing for a job in his office.

As we sat down in his office, he said, in a soft, raspy voice, "I would like you to come work for me."

"Great," I replied.

"You are a physicist, right?"

"Yes."

"Tell me the name of your PhD dissertation."

" 'An Effective Theory of Baryons and Mesons.' "

He stood up abruptly and asked, pointing at the window, "What do you think of my view?"

And so I started down the path that would eventually get me the job of commissioner, landing me inside the secret corridors of the agency charged with regulating the nuclear industry. I felt like Dorothy invited behind the curtain at Oz. Then, in another unlikely development for a guy with untested political skills and his basic idealism still intact, I became the agency's chairman.

The problem was that I wasn't the kind of leader the NRC was used to: I had no ties to the industry, no broad connections across Washington, and no political motivation other than to respect the power of nuclear technology while also being sure it is deployed safely. I knew my scientific brain could stay on top of the facts. I knew to do my homework and to work hard. But I could also be aggressive when pursuing the facts, sometimes pressing a point without being sensitive to the pride of those around me. This may have had something to do with why I eventually got run out of town. But I also think that happened because I saw things up close that I was not meant to see: an agency overwhelmed by the industry it is supposed to regulate and a political system determined to keep it that way. I saw how powerful these forces were under the generally progressive policies of the Obama administration. These concerns are even more pressing under the Trump administration, in which companies have even more power. I was willing to describe this out loud and to do something about it. And I was especially determined to speak up after the nuclear disaster at Fukushima in Japan, which happened while I was chairman of the NRC. This cataclysm was the culmination of a series of events that changed my view about nuclear power. When I started at the NRC, I gave no thought to the question of whether nuclear power could be contained. By the end, I no longer had that luxury. I know nuclear power is a failed technology. This is the story of how I came to this belief.

CONFESSIONS

OF A

ROGUE
NUCLEAR
REGULATOR

CHAPTER 1

Dr. Jaczko Goes to Washington

Born in 1970, I discovered the TV program *Cosmos* at a young age, which led to my fascination with physics. What other discipline could take a complex thing like the movement of the Apollo moon lander and describe it in a few mathematical expressions? But halfway through my five years of work toward a physics doctorate, I began to feel that the abstract world of theoretical particle physics was too removed from the real world. I wanted to use science to improve the world. And I thought there was no better place for doing that than Washington, DC.

Becoming a "sharp-elbowed political player," as one magazine later described me, was not a goal of mine. In fact after I moved to Washington, my mother took every opportunity to tell friends, colleagues, and members of Congress that I'd once sworn I would never have a job that made me wear a suit.

My ticket to Washington came in the form of a fellowship with the American Association for the Advancement of Science. For one year, alongside dozens of other scientists and engineers, I would serve as free labor for a member of Congress. So just weeks after defending my dissertation and receiving my PhD, I gave away my car and moved to an apartment in Washington that I'd rented sight unseen. It was August 1999.

My first task was to find a congressional office that would take me. Many politicians liked the idea of hiring a highly educated employee who cost nothing more than a desk, a phone, and a computer. But not everyone was willing to take a fellow. While we fellows knew a lot about some very specific branch of science or technology, many of us knew nothing about Congress and how it worked. That was certainly true of me. As a physics and philosophy major in college, I had had little time for classes in American government. Most of what I remembered about the federal government came from the *Schoolhouse Rock!* cartoons I watched as a kid.

I landed in the office of Congressman Edward J. Markey, a Democrat from Massachusetts. *Mah-key*, as his thick Boston accent sounded to my ear, was an ardent nuclear arms control expert and a passionate advocate for nuclear power plant safety. One of his first spontaneous requests for my help involved that October's Pakistani military coup and its potential impact on the Senate's impending vote on whether to ratify the United Nations' Comprehensive Nuclear-Test-Ban Treaty. Only a few weeks into my job as a congressional staffer, I was told Markey wanted to talk about the coming vote on the House floor. He asked me to write him a one-minute speech.

It had been nearly a decade since I had attempted to write anything other than dry academic papers. But knowing that Ed Markey was a funny, entertaining speaker, I knew I needed a vivid image that would bring this complex issue to life. I quickly punched out the perfect speech with the perfect metaphor: a rodeo rider on the wild bull of nuclear proliferation. With all the pride of a novice, I dashed across the office to put the speech in front of Markey's chief of staff. And then I learned, the hard way, the most basic lesson of politics: it's all local. Markey represented the dense, middle-income urban areas surrounding Boston. There were no rodeos in his district. It was the home of the Red Sox, not dirt rings with bucking bulls and leather-chapped cowboys.

Markey didn't use my speech, although in his remarks he did use the image of a bull in a slightly different way— a gesture that softened my embarrassment. I was disappointed to have struck out on my first policy statement, but I knew my tenure in his office was going to be valuable for me—and for Congressman Markey too, I hoped. Not only would I get the chance to make a difference in the world— the very reason I left academic physics—but I was going to learn from one of the best legislators in Congress. In the end, Markey and his staff taught me the strategies and tactics that were most effective in Washington, with a special focus on nuclear power issues and the Nuclear Regulatory Commission.

After eighteen months, when it was time to find a permanent job, I learned that a prominent senator was looking for a scientist staffer to help him fight another nuclear power battle. The legislator was Harry Reid of Nevada, the Senate Democratic Whip at the time.

In March 2001, I joined Senator Reid's staff after that brief meeting in his office to help him fight the Yucca Mountain project, a proposed nuclear power waste disposal site outside Las Vegas. Two years later the first of what would be several commissioner vacancies opened on the NRC. Given the role the agency played in reviewing the safety of Yucca Mountain, and the importance of that issue to the senator, Reid wanted a say in who would fill those positions. He turned to me to help identify candidates for the job.

As I dutifully started to discuss possibilities with the senator in yet another career-changing meeting in his office, he asked whether I would like to be on the commission.

"Sure," I said. And as with most conversations with Senator Reid, that was the end of the meeting. Little did I know that my nomination would be the cause of a headline-making, two-year political showdown.

The next step in my nomination, beyond excitedly telling my parents, was to wait. And wait. And wait. And wait. Nominees to commission positions become hostages for leverage in the U.S. Senate, as the confirmation process creates the opportunity for senators to fulfill other related—or unrelated—goals by placing a hold on a nomination until they get what they want. In my case, the confirmation process took two years.

Up until that point, I had been a surrogate for Senator Reid and for Congressman Markey, with very little record of my own. Since both of these legislators had been antagonists of the nuclear power industry for decades, I was guilty by association. With little to go on, the industry had to assume the worst: that my bosses' views were my views. That triggered relentless opposition from the industry and its standard-bearers in the U.S. Senate. The holds kept coming.

While I had certainly been a strong advocate for the positions Markey and Reid had taken on their respective nuclear safety issues, before my nomination I had not formed my own views about nuclear power. Technically trained, I appreciated the scientific prowess that brought the country a fleet of over a hundred nuclear reactors. But through my experience working on Capitol Hill, I had become skeptical of the ability of the nuclear power industry to properly balance its fiscal responsibility to shareholders with the demands of public safety. I was basically a nuclear power moderate: intrigued by the technology but cautious of the potential harm.

I have no doubt that had I publicly professed strong support for nuclear power and vowed to ensure the NRC did not overregulate the industry, I would have been more easily confirmed. Such a statement may have lost me the support of vocal nuclear safety advocates, but they were few in number and inured to the commission's tacitly helping the industry. Instead, I pledged to assess facts and make independent decisions, stopping short of specific pro-industry promises. The blunt message I would get over the next two years of Senate stalling was that honesty and integrity mean nothing if you are perceived to be critical of nuclear power.

Frustrated with the two years of obstruction, Reid decided to place holds on every nominee waiting to pass through the Senate's approval process—more than three hundred people—until I was confirmed. But even this muscular action—which made for great headlines in Nevada, where Reid was seen as fighting for the interests of the state—was not enough. There was one hold on my nomination he could not get released, that of Pete Domenici. The New Mexico senator was known as "Saint Pete" among nuclear proponents because of his prolific and unflinching support of the nuclear energy and nuclear weapons industries. In the mid-1990s he had made a very simple threat to the NRC: Reduce your intrusiveness by adopting more industry-friendly approaches to regulation, or your budget will be slashed. His efforts are still whispered about in the corridors of the agency's buildings in Rockville, Maryland, a ghostly memory referred to as "the near-death experience."

In a final push to win my confirmation, Reid set up a meeting between Domenici and me. I walked into the meeting prepared to defend my opposition to Yucca Mountain on Senator Reid's behalf, but when I sat down with Senator Domenici, I learned his reasons for opposing my nomination were much broader than I'd expected. He feared I would become a vocal critic of the nuclear power industry, a "mouthpiece" for antinuclear groups, as he put it. He believed that if left to its own devices, the NRC and its staff of experts would impose regulations that would destroy the nuclear power industry.

Based on the unwritten traditions of the confirmation process, senators tolerated their colleagues' holds for a certain period of time. This allowed them time to extract concessions from nominees they did not like but kept the government functioning. It was rare in those days for a commissioner to fail to be confirmed. But even after meeting with me, Domenici refused to lift his hold.

Reid did not like to lose, and so he found another way to get me onto the NRC. He convinced President George W. Bush to bypass the usual confirmation process and place me on the commission using the president's power to make appointments while Congress was in recess. The one caveat: I would be able to serve for only eighteen months unless

I was formally confirmed. To save the senators who had opposed my nomination from political embarrassment, I also had to agree to recuse myself from decisions involving Yucca Mountain for one year. This was a meaningless commitment—no substantive Yucca Mountain issues were scheduled to come before the commission in my first year—but I understood it was an important symbolic act.

And so, after two long years of political tug-of-war, Reid prevailed and I became a member of the Nuclear Regulatory Commission, leaving the legislative branch of government for the executive.

The Nuclear Regulatory Commission oversees all the commercial nuclear power plants in the United States. It is part of the family of government agencies known as independent regulatory commissions, almost all of which are known better, if they are known at all, by their acronyms: FCC, FTC, SEC, and so on. All of the commissions have a significant, if quite specific, impact on our lives as Americans.

While they regulate different industries in distinct professional fields, the independent regulatory commissions have similar structures. They are usually led by a board with five members, each of whom must be confirmed by the Senate. To ensure that each commission has, at least in theory, a diversity of views, no more than three of its members can belong to any one political party. Usually this leads to three Democrats and two Republicans or two Democrats and three Republicans. Each commissioner serves a term of five years and the terms are staggered, so one member leaves the commission every year as a new one is seated.

These agencies are designed to be independent of but not isolated from the president, whose power comes from the fact that the president chooses each board's chair. This chair wields tremendous authority, usually serving as the chief executive of the agency itself, which at NRC when I chaired it means having executive responsibility for nearly four thousand staff members and a budget of over $1 billion. Congress, however, has even greater control than the president over the independent regulatory commissions, because it oversees and funds them.

Because these regulatory commissions wield enormous power over industries like telecommunications, commercial banking, investment, and electricity, the commissioners are often the subject of intense fighting in Washington, as I learned firsthand. In the case of the NRC, powerful electric utilities strongly influence the choice of commissioners, as they depend on allies on the board for their livelihood; no nuclear power plant can operate without the agency's approval. For the past several years, this has meant that the NRC's board has been made up primarily of industry-backing commissioners. Prospective commissioners who might make safety a priority—or even dare to oppose nuclear power—don't survive the Senate confirmation process. My association with Markey and Reid would lead many in the industry to believe that I was one of the few antinuclear people to make it onto the commission in recent years.

On my first day at the NRC, my only official act was to take an oath to carry out my job faithfully; my girlfriend and later wife, Leigh Ann, held the Bible, and my family and friends watched with pride. But as I would soon learn, a host of rumors about me had already started spreading among NRC civil servants. Among the more amusing was the story that my first official act was to fire my entire staff. I did not. It was also said that a new Hummer parked in the agency's garage belonged to me. It did not. I rode public transit to the office or biked.

The NRC staffers I knew shared a dedication to the agency's core mission: protecting people and the environment from the hazards of radiation, especially from nuclear power plants. They also shared a quiet love for the NRC's Rockville headquarters. Rockville is a suburb, a strip-mall satellite of the capital, removed from the hectic political battles of Washington. For the most part, the people who made up the NRC thought of themselves as physically and mentally distanced from the political forces that shaped so many of the federal government's decisions. They weren't, though. They just felt separate, which in my mind was a good thing. A mixture of safety idealists, nuclear technology enthusiasts, and political climbers all trying to find the right answer in a mess of conflicting and uncertain possibilities, they faced an extremely challenging assignment as they engaged with the companies that owned, operated, manufactured, supported, and designed nuclear power plants.

It was in these jargon-filled direct engagements that the staff seemed most comfortable. Something was either safe or it wasn't; the numbers would tell. But these well-intentioned technical wizards sometimes had a hard time finding the right words to communicate this to the people who lived near the nation's nuclear plants. For this reason, the agency sometimes came across as a collection of uncaring, robotic bureaucrats. I knew the staff was anything but, and I was determined to make this clear to the outside world.

That being said, it was also true that those who wanted to climb the ladder to senior management had to become closely (arguably too closely) acquainted with the agency's most significant associates: the companies that compose the nuclear industry.

Although people talk about the nuclear power industry as if it were a monolith, nuclear power is produced by many different companies in many different sectors of the economy. Some of their names are familiar: General Electric, Westinghouse, Toshiba. Most of them make products, plants, and services that create all types of electricity, not just from nuclear power, using a combination of traditional and renewable energy resources.

What all of these disparate electricity producers, suppliers, and distributors have in common is membership in the Nuclear Energy Institute, the lobbying organization representing the industry's interests. NEI has the unenviable task of corralling these powerful companies, each with its own goals. But when it comes to influencing laws and regulations, NEI members have a history of acting as one. This solidarity gives them tremendous influence with Congress.

NEI also has a huge impact on the decisions of the Nuclear Regulatory Commission; this is where member companies recoup their million-dollar membership fees. Killing regulations, or even modifying them slightly, can produce savings of millions of dollars per year in operating costs, equipment purchases, and technical analysis. With millions to spend and a unified message, NEI shapes every NRC regulation, guidance, and policy. In some instances,

NEI works through formal channels, commenting on documents produced for the public. In others, it exerts its power through informal meetings with commissioners. In any given month, I could be visited by as many representatives of the industry as I would be by public interest groups across my entire seven and a half years on the commission.

A typical visit from a representative of NEI or a utility company would start at the middle manager level and end with the commissioners. That way, if NEI heard troubling news from midlevel staff, they could raise the issue with one or more friendly commissioners, and actions would be taken. I saw this happen all the time, even though staff members were repeatedly told to not take direction from commissioners or industry executives. The commission's role was merely to set policy through formal votes; only the chairman had the authority to carry out that policy. Having five bosses telling some staffer what to do often led to chaos and paralysis. But if the industry's goal was to prevent regulation, then chaos and paralysis were a plus. This was one of the many problems caused by the industry's influence over the agency. As a commissioner, there was little I could do to correct any of them. But after four years on the commission, my opportunity came when Barack Obama became the president of the United States.

Obama's election created massive upheaval in Washington. Thousands of jobs would change from Republican hands to Democratic. Tradition dictated that a Democratic president would choose a Democrat to serve as chairman of the NRC, either a new commissioner to fill a vacancy or an existing commissioner. At the time of Obama's election, I was the only Democratic commissioner; the Senate was waiting on the outcome of the election to fill the vacancies. Fortunately for me, Senator Reid was not intimidated by the nuclear industry proponents' efforts to stop me from being named chairman, but it took even him—by that time the Senate majority leader—over six months to get his way.

Still, before Obama could officially name me chairman, Reid told me I would need to meet with Rahm Emanuel, the president's chief of staff. I was told little about the purpose of the meeting, but I assumed Rahm would want to hear about my vision for the agency and what I planned to accomplish. I dutifully dusted off my memos about the improvements to nuclear safety that needed to be made, and within a day I was on my way to the White House.

As I passed through the many black gates that regulate access to the grounds, I went over the most important points that I wanted to discuss: the agency needed to focus on a number of unresolved safety issues; the agency had to be more aggressive in dealing with the industry; the agency should undo the damage caused by my predecessor's heavy focus on industry protection. I remembered hearing Senator Obama's words on the radio one day during the presidential race, criticizing his opponent's economic policies as "a philosophy that says even commonsense regulations are unnecessary and unwise." For four years as a commissioner, I had been accused by the industry and some members of Congress of being an extremist and an uncooperative colleague because I had tried to carry out commonsense regulations. My agenda, I believed, was in sync with the new president's vision of a government committed to doing what was in the best interests of the American people.

To the U.S. Marines guarding the outer door of the White

House and the receptionists controlling the inner corridors, I must have looked like every other young, ambitious person excited at the prospect of making a difference. I was led to the office of Jim Messina, one of Rahm's deputies and a former Hill staffer like myself. Jim was seated behind his desk. His eyes were narrow as if he were squinting, and he seemed determined to let me know that I should not underestimate him. I started to sense this meeting was not going to be the friendly welcome I was expecting. Something surprising was going to happen; I just couldn't figure out what.

I sat across from Jim, my excitement at the opportunity to demonstrate I was ready for the challenge of leading the NRC causing me to ramble on about arcane nuclear safety statistics. Jim spoke a bit about the president's priorities around health care and climate change as we continued to wait for Rahm.

Suddenly Rahm appeared. When I stood to shake his hand, he looked me in the eye, then sat in a small chair against the wall, his body so tense it was as if he might jump up at any moment and bolt from the room. As he was seated to my right, I was in the uncomfortable position of sitting sideways to him. I could either look at Rahm and ignore Jim or look at Jim and disrespect Rahm. It was clear Rahm knew what he was doing in picking that seat, taking every opportunity to make me feel ill at ease. I decided I'd look straight at him. For the next few minutes, my gaze never left his face.

"You are a fucking asshole and nobody likes you," he declared. "If we make you chairman, everyone at the NRC is going to quit. No one wants you to be there. I'm only telling you what no one else will tell you."

This was not the opening I had anticipated. All I could think was, *I guess I won't be telling my parents about this meeting.*

"I personally don't care about nuclear power," Rahm continued, a bit of spit flying from his mouth, "but the president wants to address climate change, and he needs to have nuclear power as part of that program. So he needs the NRC to do its job—and I don't think you can do the job." Just as I was about to declare my full support for all the president's priorities, he cut me off. "Don't think this is just coming from me. I spoke with the president this morning and he asked me again, 'Why am I going to make this guy chairman when no one wants him to be chairman?' I said to the president, 'Because Harry Reid is the best Senate majority leader there is and Senator Reid wants him to be chairman.' So the president is going to do this for Harry Reid."

I took a breath. Apparently the president either did not agree that we needed tough nuclear safety regulations or didn't think I was capable of doing the job. When I entered the White House that day, I had expected to leave proud of my new opportunity. Now I was just hoping the snipers on the White House grounds wouldn't take aim at me.

"Being chairman is a very important job," Rahm went on. "I don't expect you to make problems for the president. Do you understand that? You work for the president and you better not fuck this up."

"I understand that I would work for the president," I blurted out. But not even my profession of allegiance would stop the insults.

"The president really does not want to do this, but he is

going to sign the paperwork to make you chairman," Rahm continued. "I do not expect to hear about any problems from the NRC."

As quickly as he entered, he left. And whatever confidence I had went right out the door with him. Stunned, I turned to look at Jim, hoping for some explanation. He offered me no sympathy and reiterated the major points: "Health care and energy are the president's two most important issues. And nuclear power is crucial to his energy program. We don't need any distractions from that basic goal. So don't fuck it up."

I took this to mean that I shouldn't be too hard on the industry because the president needed its support to address his climate change goals. As the blood returned to my face, so did a tiny part of my pride. I uttered the words I wish I had said to Rahm: "I am not an asshole. The NRC has been under poor leadership for the past few years and I'm going to need to clean it up." Jim showed no strong reaction to my words.

Before the meeting ended, Jim and I worked out the details of how I would communicate with the White House in the future. Fearing Rahm would be my point of contact, I was relieved to learn I would report to Jim. With that, I collected my never-opened briefcase, tried to hide the shaking in my hands, and left. As my shock over the ferocity of Rahm's tirade lessened, I realized I had actually gotten the job.

Days later I assembled the senior managers whom I would lead as chairman. It was clear I was not the kind of boss they were used to. I was younger than most of the people who would report to me, and I lacked experience leading a large organization. I was hardly their first choice for the job.

While my political allies, most notably Reid, were powerful, they were small in number and generally seen as antagonistic to the industry. And although I had already spent more than four years at the agency, I had kept my distance from industry leaders. I knew them and they knew me, but I believed it would be easier to make objective safety decisions if I didn't get too friendly with them. As a commissioner, my reserve was not such a problem; the industry trusted that they had enough support on the commission. The worst I could do was publicly expose embarrassing problems. I was unlikely to successfully promote a more aggressive nuclear safety agenda.

As chairman, however, I would now be the chief executive and chief spokesperson for the agency. I was responsible for setting its agenda, leading its nearly four thousand employees, developing and defending its budget, and speaking to Congress, the media, and the public about the safety of the nation's nuclear power plants. The industry, which had lobbied against my becoming chairman, was in a weakened position. I understood the cutthroat nature of Washington politics, so I held no grudge. But I also had no obligation to the industry; I was free of political constraints. I was also short of allies on the commission to help me fix the problems I had found. This made for a promising but problematic start to my tenure.

One of the first things I did was signal to people at the agency that I had their back. I held lunches, hosted open houses, and invited employees to join me in meetings with top industry executives. My comfort with the issues and my deepening connection with the NRC staff undoubtedly raised concern among industry leaders that I might actually be effective at my job. But I was not without faults and weaknesses, limitations that would come back to haunt me when the industry mob came after my head. At times of high stress or extreme frustration, I could raise my voice—never with the intention to intimidate but in a way that could leave that impression. I was passionate, but at times that made me move too fast to think about how I made people feel. In Washington parlance, my IQ sometimes outweighed my EQ. But these weaknesses would not have mattered if I hadn't committed the much greater sin of becoming a threat to the industry's fortunes.

The conflict evolved slowly. My first efforts were focused on getting into the nuclear oversight weeds to fix highly complex technical issues. Even the powerful nuclear lobby could not generate much horror in any but their most ardent supporters about my meddling with these esoteric but vital details of public health and safety. These were hardly the kinds of issues that would spill onto the pages of major newspapers or make the evening news.

Then bigger issues came along. The first arose when I pushed to make good on the president's promise to end the program to store nuclear waste in Nevada. The administration had bungled the effort to close down the Yucca Mountain project, so I stepped in, using the full authority of my office to finish the job. My actions created a rift between my colleagues and me (as a majority of them did not want to see the project end) and alerted the industry's congressional supporters to the powers I could use as chairman.

But the biggest battle began after almost two years as

chairman when the worst nuclear accident since the 1986 Chernobyl disaster happened at a nuclear power plant thousands of miles away in Japan. Before the Fukushima accident, there were many nuclear professionals in the United States and Japan who believed there would never be another significant nuclear accident. Unfortunately, they were wrong, and for a very simple reason: no one can design a safety system that will work perfectly. Reactor design is inherently unsafe because a nuclear plant's power—if left unchecked—is sufficient to cause a massive release of radiation. So nuclear power plant accidents *will* happen. Not every day. Not every decade. Not predictably. But they will happen nonetheless.

The designers of nuclear facilities would not agree that accidents are inevitable. When building their safety backups, they essentially say, "Whatever you need, double or triple it." If it takes one pump to move water during an accident, for example, then put in another pump somewhere in the plant. However, this fail-safe setup only reduces the chance of an accident; it does not eliminate it. What if a failure disables both pumps simultaneously? And what about the problems that no engineer, scientist, or safety regulator can foresee? No amount of planning can prepare a plant for every situation. Every disaster makes its own rules—and humans cannot learn them in advance. Who would have thought a tsunami would cause a nuclear disaster in Japan?

Uncertainty about when an accident will happen is exactly why the industry makes the argument for doing nothing. "Why spend billions of dollars to prevent something that might not happen for thousands of years, if at all?" they say. But the accident at the Fukushima plant is a rebuttal to that argument: despite decades of advances in safety systems, reactor physics knowledge, and nuclear plant operator performance, a catastrophic accident shocked most of the world simply by happening. Maybe another accident won't happen for thousands of years. Or maybe it will happen tomorrow.

Many tried to dismiss Fukushima as a result of Japanese unwillingness to challenge authority. Their engineers simply didn't push back against the norms that stand in the way of safety, people said. But that same obeisance to the powerful is exactly what I saw at home in the NRC.

When I realized how flawed the safety technology was—not just in Japan but at U.S. nuclear facilities—I decided I would do everything I could to fix it. My determination set up a major conflict between my fellow commissioners and me. Following the Fukushima accident they appeared to me most concerned with preventing the agency from inflicting pain on an industry now struggling to respond to a major nuclear power plant accident in a country far away.

The Fukushima disaster forced one hundred thousand Japanese from their homes. Brave workers were exposed to high levels of radiation as they tried to arrest the damage. Japan's economy was rocked. But American politicians had long ago been led to believe that these kinds of calamities were no longer possible. And so pressure was placed on the agency—even after the disaster—to do just enough to say safety was taken care of, but not so much that it forced the industry to make meaningful changes. From my prime seat at the most significant contest over the future of nuclear power, I saw the industry and its allies continue to try to thwart even the most basic and commonsense safety reforms.

In hindsight, the Fukushima incident revealed what has long been the sad truth about nuclear safety: the nuclear power industry has developed too much control over the NRC and Congress. In the aftermath of the accident, I found myself moving from my role as a scientist impressed by nuclear power to a fierce nuclear safety advocate. I now believe that nuclear power is more hazardous than it is worth. Because the industry relies too much on controlling its own regulation, the continued use of nuclear power will lead to catastrophe in this country or somewhere else in the world. That is a truth we all must confront.

CHAPTER 2

Forget and Repeat:
A Brief but Necessary
History of Accidents

The Fukushima accident in Japan was not the first accident to belie the promise of nuclear power. In its early years, the commercial nuclear industry had only a limited understanding of the operation, science, and engineering of actual power plants. This ignorance led to the first major nuclear power plant accident, just outside Harrisburg, the state capital of Pennsylvania, in 1979. Three Mile Island prompted a flurry of reforms and a pile of promises that the public would be protected from future nuclear calamities. Through the mid-1980s, it appeared these promises were being kept; construction on new plants slowly resumed without major accidents. Then, suddenly, strange radiation measurements were detected in Sweden. Governments in Europe and throughout the world soon learned that a disaster had occurred at the Chernobyl nuclear power plant in the Soviet Union.

Like a developing photograph in a bath of chemicals, the reality of nuclear power was starting to become clear. One nuclear accident was an oversight, a mistake, an aberration. Two nuclear accidents hinted at a serious problem with the technology. A third would cement the conclusion that nuclear power plants were simply going to have accidents on a relatively consistent schedule. After Three Mile Island, after Chernobyl, that third accident nearly occurred in 2002 at the troubled Davis-Besse nuclear power plant in Ohio.

The problem is that with each new accident, all the people in charge of nuclear safety seemed to revert to the belief that this one would be the last one. As chairman of the NRC I battled nearly every day against this instinct to believe the worst was over. You can prepare for the next accident only if you can get all the players to admit that a next one is coming, even if when and where are impossible to predict. Before Fukushima, too many people I encountered simply did not believe the next one would ever come. Their view was not surprising; accidents are rare, and Chernobyl and Three Mile Island had happened decades earlier. Yet I continued to believe I could challenge this complacency. I seized one opportunity just after I became chairman.

Four days after President Obama tapped me to lead the commission, I spoke at a conference organized by the North American Young Generation in Nuclear, an industry group for professionals entering the field as plant operators, designers of reactors, or academic experts in nuclear technology. As I looked out at the crowd, it dawned on me that many of these people had never lived through a nuclear power plant accident. Even I had been only nine years old when Three Mile Island occurred; when Chernobyl happened I was a teenager more worried about surviving my freshman year of high school than about nuclear disaster. The people I was speaking to were even younger. I wondered how they had experienced these seminal events.

Being a scientist, I decided to conduct an experiment. I asked everyone in the audience to stand if they were born after 1979, the year of Three Mile Island. Nearly everyone stood. After they sat down, I asked them to stand if they were born after 1986, the date of the Chernobyl accident. Once again, nearly everyone stood.

These industry-defining accidents have become dry case studies taught in college classes. The next generation of American nuclear power professionals has never experienced the confusion of a nuclear accident as it is happening. And so it's essential that we remember and teach the lessons of Three Mile Island and Chernobyl, for reviewing these accidents shows common themes of missed opportunities, human failings, and technological overconfidence. No amount of forgetting can change these simple facts.

The March 1979 accident at the Three Mile Island nuclear power plant in Pennsylvania seems almost like something out of a science fiction horror film. The cover of *Time* magazine captured the national mood of chaos, confusion, and fear, the emergency-red phrase "Nuclear Nightmare" splashed across the dark black cooling towers of the plant. There was no live-streamed video as there would be after the Fukushima accident, but the public could imagine the scene inside the reactor. Just twelve days before the accident, *The China Syndrome,* a feature film starring Jane Fonda and Michael Douglas as reporters who uncover a major incident at a nuclear plant, had been released. Perhaps the hundreds of journalists gathered outside Harrisburg believed they too would land such a story.

It started on March 28 at around 4:00 a.m., when a water pump stopped working. The failed pump affected the steam generators, large cylinders filled with many tiny metal tubes that help turn hot water from the nuclear engine into steam so that the turbines can create electricity. When the flow of water was cut off, this massive heat exchange stopped working, creating the conditions for a serious accident.

The reactor engine was immediately turned off. But so long as the reactor fuel remained hot (which it would for quite some time), its natural radioactive decay would continue, producing enough heat (called "decay heat") to melt through the metal containers enclosing the reactor fuel. (This same problem would later affect the Fukushima plant.)

The failure of the main feedwater pump was not in and of itself a serious crisis. But the systems responsible for removing the decay heat—and the people operating those systems—did not respond correctly. As the reactor shut down, the closed cooling system suddenly no longer had anywhere to deposit its energy. This caused a significant spike in pressure in the pipes circulating water to cool the reactor.

Plants of this type are outfitted with a large tank of water designed to regulate this pressure; it's called a pressurizer. Like a bob on a fishing line, the pressurizer water level rises and falls to keep the pressure consistent. When it gets too high, a valve opens to release some of that pressure. During the initial phase of the accident, this safety valve did something it wasn't supposed to do: it stayed open after the pressure had been relieved. Operators can fix a stuck pilot-operated relief valve, as this pesky component is called. But the people running the plant were let down by their instruments. The control panel, with all its lights, knobs, and switches, told them the valve had closed.

The open valve allowed essential water to pour out of the pressurizer, draining the reactor vessel, exposing the nuclear fuel to air. These hot fuel rods now lacked the necessary cooling to keep from melting.

Nuclear power plant operators are taught to worry about a serious condition in which the pressurizer, which balances pressure through a combination of gas and water, goes "solid," meaning that all the gas has gone, leaving only water behind. This scenario leaves the operators unable to regulate the pressure. It is as if the fishing bob has gotten caught in some branches and cannot allow the fishing line to go up and down.

Seeing the pressurizer appear to go solid—as they were taught to expect—the operators reduced the water in the reactor cooling system. This made the reactor fuel even hotter. As the pressure dropped throughout the system, the immense pumps that circulate water through the plant began to vibrate fiercely. To protect the pumps, the operators turned them on and off, further reducing the heat removal capability of the limited amount of water left in the reactor vessel. The fuel began to melt, releasing a burst of radioactive material into the containment structure. By evening the reactor's normal cooling had been restored, but the damage was done.

Outside the walls of the Three Mile Island plant, the confusion was just beginning. The first signal that something serious might be happening came when a general emergency (the highest level of safety alert) was declared around 7:00 a.m. Because of ineffective communication, however, this alert did not reach the NRC's regional staff outside Philadelphia for another forty-five minutes. Contacting government officials—even in an emergency—is never easy, and this was before cell phones and text messaging. Since the NRC rarely required power plants to notify the agency about less significant issues, these communication challenges were only now becoming apparent. It would take a few more hours before the White House learned about the situation.

Nothing about this communication failure is unique. As I learned in the wake of the Fukushima accident, crises on this scale are often characterized by incoherent communication and conflicting information. Both the Three Mile Island and Fukushima disasters featured contradictory assessments of the state of the reactor, a limited appreciation of the fact that the damage to the reactor had occurred very early, and rapidly changing statements from elected officials.

To the public, these statements can appear to suggest prevarication or incompetence. But when government officials—imperfect human beings like everyone else—try to make sense of the complicated physics of a nuclear reactor accident, they will invariably make mistakes in communication.

There is also a subtle bias against believing the worst is happening. The Kemeny Commission, which investigated the Three Mile Island accident, would capture this sentiment exactly when Gary Miller, the station manager, admitted, "I don't believe in my mind I really believed the core had been totally uncovered, or uncovered to a substantial degree at that time." Knowing what is happening and believing what is happening can be two very different things.

After a general emergency was declared at the Three Mile Island plant, the governor of Pennsylvania, Dick Thornburgh, chose not to execute an evacuation. Although state officials are responsible for such decisions, they rarely have the background in nuclear technology to accurately assess the situation and instead rely on experts at the plant or the NRC, who are also scrambling to understand what is going on. Of course, communication between these disparate groups is never perfect. Elected officials in Harrisburg received updates from the press instead of the plant. In a day when we're accustomed to getting our news from 140-word tweets, such speed—if not accuracy—is to be expected. But for the press to have had more reliable information sooner than government officials in the era of rotary phones, teletypes, and just three broadcast networks seems shocking today.

As operators worked to restore a normal cooling cycle, the uncertainty over the condition of the reactor and the possibility for further damage created challenges for officials attempting to formulate a plan. On the first day of the accident, a small hydrogen explosion occurred in the reactor—a clear indication that the reactor fuel had already been severely damaged. But staffers in the plant dismissed the event, believing the public was not at risk. The only major radiation readings were coming from inside the buildings that housed the reactor and the major safety and operational equipment. There were not yet any significant readings of off-site radiation.

Nonetheless nuclear analysts in federal and state government began to consider calling for an evacuation if the reactor deteriorated further. These discussions, based partially on federal government officials' misinterpretation of information coming from the plant, led some NRC staff to recommend that state emergency management officials evacuate those living near the plant. But NRC personnel at the reactor and other workers there disagreed. Eventually these conflicting messages were reconciled and Governor Thornburgh issued a shelter-in-place order two days after the accident began, instructing people living within ten miles of the plant to stay in their homes and not venture outdoors. (The plan actually called for a five-mile radius; the increase to ten miles was due to miscommunication.)

Within a few hours another conversation between Thornburgh and the chairman of the NRC led the governor to finally issue an evacuation order for pregnant women and school-age children living within five miles of the plant. This order left the public confused and frightened by a government that appeared confused and indecisive.

The evacuation order would be the last public safety recommendation; ironically it was issued after the danger had passed. Uncertain about the extent of the damage to the reactor, however, NRC technical experts grew concerned about the potential for a hydrogen bubble to form and ignite, potentially releasing even higher levels of radiation into the environment. The lack of precise details about the amount of hydrogen in the reactor, and its ability to explode, led to a series of highly public and technical debates. After a number of stressful days, concluding with a visit from President Jimmy Carter to Three Mile Island, the NRC staff determined that a hydrogen bubble was unlikely to form.

The accident was over, but more than ten years would pass before the plant would be cleaned up. Over $1 billion was spent to recover and dispose of the damaged reactor fuel. The nation may have avoided a nuclear catastrophe, but the costs were high—and Americans had lost confidence in nuclear power.

The alarms that rang out at Three Mile Island would reverberate for years to come. The accident raised serious concerns about the effectiveness of both the companies operating nuclear plants and the federal and state agencies looking over their shoulder. Several highly critical and consistent post-accident studies captured these concerns, with findings that would lead to significant changes in the structure of government agencies and the operation of utilities.

The Three Mile Island accident almost led to the abolition of the NRC. Troubled by the agency's lack of a clear leader in the crisis, the president and Congress considered replacing the NRC with a more traditional government agency led by a single individual. Eventually a compromise was reached: the NRC would remain, but the law would be changed to ensure that its chairman was unequivocally in charge during an emergency.

The Three Mile Island accident exposed serious weaknesses in the control rooms, communication and safety systems, and operations of nuclear power plants, leading the NRC to add or modify countless regulations to address these shortcomings. Control room layouts, emergency procedures, and operations practices were changed. More alerts and information panels dotted the control boards. Operators would now train more extensively. Additional inspectors from the NRC established a permanent presence at each of the nation's nuclear power plant sites.

The industry was shaken. Between Three Mile Island and the bleak economic conditions at the end of the Carter administration, proposals for over one hundred new nuclear power plants were canceled. And the companies that owned and operated America's nuclear power plants recognized that a single accident at any plant could provoke enough concern that all plants could face more regulation or even a forced shutdown. This shared risk created a strong bond. Together the owners established a new organization to try to protect themselves: the Institute of Nuclear Power Operations, which worked to ensure that this accident would be the last in the United States.

But another nuclear accident took place seven years later.

A Soviet plant was conducting an experiment to improve the reactor's ability to handle an emergency shutdown. To do this, the operators disabled a number of the plant's safety systems and alarms. They then unintentionally put the plant in a very unstable condition. Less than a minute after the test began, two large explosions occurred. The first destroyed the reactor core, causing a massive burst of energy that tore the roof off the building. The second ejected hot and highly radioactive fragments of the reactor core into the air. With the remaining portion of the core now fully exposed, the reactor began to burn. Over the next several weeks, emergency crews worked to stop the massive release of radiation, dropping special materials on the wrecked reactor core to halt the nuclear fission and stop the fires.

The Iron Curtain—that figurative, ideological dividing line between the Soviet sphere of influence and the West—was no match for winds blowing contamination across Europe. Within two days of the incident the radiation released in Chernobyl was picked up on the soles of a worker's shoes during a routine scan at the Forsmark nuclear power plant in Sweden. Turning to radiation monitors that had been set up throughout Sweden, scientists were able to trace the radioactive contamination to Ukraine. (Radioactive elements give off unique signatures, making nuclear forensics a very precise science.) Only then did the Soviet Union acknowledge the accident.

Still, Chernobyl got little attention. For one thing, the accident happened within the cloistered Soviet Union, so the first detailed account of what took place came to the West only four long months later. The Soviet Union's nuclear plants were also technologically and operationally different from most in the West, which meant that what went wrong at Chernobyl did not exactly apply elsewhere. While water, for example, performs many of the operational and safety functions in American reactor systems, the Chernobyl reactor relied on graphite, which significantly increased the accident's radiation contamination. What's more, the accident read like a handbook of everything not to do when operating a nuclear power plant. Even the most ardent nuclear opponents would have had a hard time believing the people who controlled nuclear plants in the West would be so careless. And so political, social, and scientific isolation created a perfect excuse to dismiss the Chernobyl accident as irrelevant to a newly recovering nuclear industry in the United States.

The Chernobyl accident remains to this day the worst in the history of commercial nuclear power. Like the Three Mile Island accident before it and the Fukushima disaster after it, Chernobyl required the evacuation of people living around the plant. But unlike those other accidents, people at Chernobyl died. Because of the magnitude of the explosions and the physical damage to the plant, first responders were exposed to significant radiation. Thirty emergency workers died shortly after the accident, and another 106 people suffered from acute radiation sickness. These survivors experienced a variety of debilitating health effects, including cataracts and skin damage. As time passes, the health authorities tracking these affected workers find it more and more difficult to know which diseases that developed in later years are due to earlier exposure to radiation. But it is likely that at least some of these survivors will die, or have already died, from ailments directly related—although not verifiably so—to their exposure to radiation from the accident.

Five million people living in the path of the contamination plume in what are now the independent states of Belarus, Ukraine, and Russia were also exposed to radiation from the accident. Thousands of children and young adults developed thyroid cancer, most of them from drinking milk contaminated with a radioactive form of iodine; had the government properly restricted access to the contaminated milk, these cancers could have been prevented. It is impossible to know just how many people in the region—not to mention elsewhere in Europe, where lower but still significant levels of radiation were found—suffered other illnesses as a result of Chernobyl. Scientific experts, concerned citizens, and government officials have widely varying assessments of the damage done.

And yet, as I mentioned earlier, the Chernobyl disaster was marginalized by the American nuclear industry as a foreign accident with a foreign cause. Despite coming only seven years after the incident at Three Mile Island, which proved that accidents could happen at home, Chernobyl was not used as a learning opportunity. The NRC's final assessment of the disaster found that no changes should be required by American plants. The world's most significant commercial nuclear reactor accident would have no discernible impact on the nuclear industry in the United States.

How different things would be fifteen years later, when a plant in Ohio narrowly avoided a major accident.

Before Fukushima, the most prominent nuclear incident in recent times took place at the Davis-Besse nuclear power plant near Toledo, Ohio.

As so often happens, Davis-Besse's problem had begun years before it was finally discovered. The designers of the first wave of nuclear plants had limited experience with the metals and other materials used to build these structures, so some of their choices turned out to perform worse than expected in the high heat, harsh radiation, and extreme chemical environment of nuclear reactors. One of these metals,

Alloy 600, had by the late 1980s begun to crack. The NRC's initial assessment concluded that any failing plant parts made of Alloy 600 would exhibit obvious signs of decay; fluids, for example, would start to leak through cracks. Still, the cracks would grow slowly, many in the nuclear world believed, allowing ample time to act before any significant problems could occur.

By the late 1990s, as cracks in Alloy 600 continued to be found at plants around the country, the Nuclear Energy Institute, the lobbying arm of the industry, insisted that these cracks would not affect safety. Not completely satisfied with this conclusion, the NRC decided to analyze the suspect material's potential for failure and to do closer inspections of the power plants that might be affected. Not surprisingly, the inspections turned up more problems. In 2001 the agency finally issued a direct requirement obligating nuclear power plants to address the issue—more than a decade after the problem was first identified.

Throughout that decade each additional probe into Alloy 600 conditions had identified new physical evidence suggesting the problem was worse than the models and many nuclear safety professionals had predicted. This is one of the more important implications of Davis-Besse: despite decades spent evaluating nuclear reactors, we can always discover new problems that surprise us. This challenges the idea that professionals can ever really know for sure what's safe when it comes to a nuclear plant.

After the NRC's first formal notice about the vulnerability of Alloy 600, plants responded in a variety of ways. Some made modifications quickly; others asked for more time. This second approach is typical in the nuclear industry. No issue ever appears to be pressing because there is a mistaken belief that early warnings inside the plants themselves will always preface a major incident. Leaks will appear well before pipes ever break. Inspections will catch cracks before they grow big enough to affect the performance of vital safety equipment. Fires will be caught and extinguished before they can spread. The operators of the Davis-Besse plant shared this complacency.

The issue at Davis-Besse started with the reactor pressure vessel head, which had parts made of Alloy 600. This large steel lid caps the container housing the reactor fuel, making it one of the most important barriers keeping radioactive material out of the environment. Like most barriers in a nuclear plant, the vessel head has openings to allow equipment to access the reactor fuel and measure the status of the reactor engine. One of these penetrations that dot the top of the lid like a series of chimneys was severely corroded.

The cause of the corrosion was boric acid, which had leaked through cracks in the Alloy 600. (Boric acid is added to the water used to cool the reactor to help control the nuclear fission process.) The corrosion made the surface of the metal look like popcorn—not a difficult sign to miss. Indeed the signs of boric acid corrosion are so unmissable that the NRC was confident operators would notice any prospective problem long before it posed a hazard. But at Davis-Besse, if anyone noticed, no one said a word.

Earlier in 2001 the NRC had asked all plants to send data on the conditions of parts made from Alloy 600 and the ability of inspection programs to identify cracks long before they became a cause for alarm. This information was due in December. But Davis-Besse delayed responding to the agency's request. The operators planned to gather the information the following spring, when the plant would shut down to perform routine maintenance.

Worried about the risk of waiting until spring, the NRC ordered Davis-Besse to stop operations. Faced with this threat, the plant shut down in February 2002, several months later than the agency's technical experts believed appropriate and not long before when the plant had planned to shut down anyway. Subsequent inspections revealed extensive damage: the six-inch steel vessel head had corroded away completely. During the inspection, the chimney-like protrusion where the leak originated toppled over like a domino, hitting the one next to it. The only remaining barrier to the reactor was a thin piece of steel not designed to hold back the pressures that would come during operation. Had Davis-Besse been in operation, a significant accident would likely have occurred. Pressurized water from the reactor vessel would have shot out through the hole, damaging the safety systems nearby.

The incident was a tremendous embarrassment to the industry and the agency. Warning sign after warning sign from inspection after inspection had indicated that there was a leak in the reactor pressure vessel head, yet neither the NRC nor the plant owner took action. While the Three Mile Island accident was the result of a minor equipment malfunction followed by human error, the problem at Davis-Besse was in some ways much more serious. The damage to the reactor vessel was so significant that had the thin steel liner failed, there would have been no easy remedy, no matter what the operators did.

There followed the usual round of hand-wringing, re-port writing, and penance serving. The Davis-Besse plant owners received a record fine of $5.5 million from the agency and $28 million from the Department of Justice, a pittance compared to the cost of the accident that would likely have occurred. At a time when some nuclear plants were generating profits of nearly $1 million a day, this was hardly a significant penalty. No senior executives were held responsible, although two lower-level employees and a contractor were charged with lying to the agency about the reactor's condition. The employees were convicted, sen-tenced to probation, and fined.

The NRC launched a massive effort, the Davis-Besse Lessons Learned Task Force, to try to prevent this kind of systematic human failure from happening again. The pro-gram lasted for more than a decade, well into the time I served on the commission.

It is difficult to prevent the kinds of systematic failures that characterized the Davis-Besse accident, especially since the false information provided by the people crimi-nally charged made it harder to identify what actually went wrong. And as an attorney once told me, difficult cases make for bad laws.

This leaves the nuclear power industry in a precarious situation. Nuclear accidents both major and minor will happen again. They always do, because safeguards inevita-bly have holes. How big are those holes? We don't know. What Three Mile Island, Chernobyl, and Davis-Besse show is that no one can be sure how big the holes are until someone falls through them. Yet the extensive reviews that followed these three incidents also pointed out facts that were obvious in hindsight, facts that suggest the people designing the safety precautions should have recognized the perils from the start. But *We should have known better* is never an acceptable explanation after a nuclear power plant accident.

CHAPTER 3

The Burning Issue: The Battle to Prevent Nuclear Fires

Fire is one of the biggest hazards inside a nuclear plant. With duplicate and triplicate safety systems throughout, the worst dangers come from events that can take out all these systems at one time—a "common cause failure" in industry jargon. A plant's maze of hallways and passageways provides an easy environment for heat and flame to sweep through, causing potentially unfixable damage to safety systems. The flames' most vulnerable targets are the data and power cables that supply information about vital plant systems and make those systems work.

In the late 1990s, calculation after calculation by modern computer models confirmed that fire brought the most significant risk of complete breakdown at many nuclear power plants. Yet the industry and the regulators were slow to grasp the importance of these models, so slow that by the time I became NRC chairman in 2009 this issue was still unresolved. My attempt to improve the ability of nuclear power plants to deal with fires turned into a drama featuring industry foot-dragging, obfuscation, and downright resistance.

———

I find sitting at a desk all day confining. So whenever I saw on my schedule a meeting with the agency's fire protection safety experts, I got excited. They would bring props and do show-and-tells with burned stuff. If I was especially lucky, they would take me to a nearby facility where they experiment with fire.

Just a thirty-minute drive from my office, the Germantown offices of the National Institute of Standards and Technology has an impressive experimental facility in which you can burn stuff. Lots of stuff. Really, really hot stuff. But this is no arsonists' training ground. As you enter the bland concrete structure, one of many buildings that populate the NIST campus, you pass a monument to the technical expertise of the people who work here: a twisted steel beam from the World Trade Center in New York. The people inside this building had been called upon to create the computer model showing how the World Trade Center buildings had crumbled in on themselves following the terrorist attack on September 11, 2001.

These technicians—with their still childlike scientific enthusiasm—had the job of refining a new generation of computer tools that the NRC was going to need to finally meet its long-standing fire safety responsibility. It was important work. And I found these visits to NIST a diversion from the stress of the job and a time to feel like a scientist again.

On my first visit, the fire engineers set up a very simple experiment, based on the first lesson every child learns about a fire: If you are caught in a fire, get down on the ground and crawl. From a more sophisticated perspective this experiment was a tool to validate computer models that simulated fires in closed rooms, a common feature inside nuclear power plants. The fire scientists lit several bales of hay and told me a clear line would quickly separate the fire's smoke from the breathable air. The smoke streaming from the hay hardly looked orderly as it rushed up to the exhaust hood at the top of the room. But just as the assembled experts predicted, a crisp, flat plane formed, with muddy, caustic smoke above and hot, clear air below.

On other visits to the burn facility, I observed more tests to refine the computer models that nuclear power plants study to protect themselves against fires. After seeing a few more experiments, I was convinced not just about the way smoke behaves during a fire but about the integrity and expertise of the federal employees I met there. I trusted them to challenge the industry lobbyists and commissioners who were stalling instead of protecting nuclear plants from catastrophic-accident-inducing fires. Now all I had to do was figure out how to convince the rest of an intimidated bureaucracy to join us in this fight.

My interest in fire safety was sparked years earlier, when I learned about the 1975 fire at the Browns Ferry nuclear power plant in Tennessee. That fire grew out of routine maintenance work on the buildings that housed the reactor. Despite their formidable size, the containment structures of many nuclear power plants, designed to corral dangerous radiation in the event of an accident, are punctured by vents and ducts. These penetration points are the weak spots that can undermine an otherwise airtight containment shell. A leak in one of these areas is a significant problem.

Confessions of a Rogue Nuclear Regulator

On March 22, 1975, just a few months after the Nuclear Regulatory Commission officially became the nation's nuclear safety watchdog, workers at the Browns Ferry plant were doing repairs to some of these penetration seals. They were working in the cable spreading room, a location near the control room that gathers together all the cables that connect to the plant's various instruments, motors, pumps, and other mechanical and electronic equipment. A fire in this room could disable all those faraway components without even entering those parts of the plant. The Browns Ferry fire did just that.

The workers were searching for a possible leak in the walls separating the reactor from the public. To determine the location of a draft—which could serve as an escape route for dangerous radioactive material—a technician held a candle up to places where there might be holes and watched to see if the flame wiggled in the slight breeze of outward-flowing air. While performing this low-tech examination, the technician held the candle too close to a nearby cable; its insulation started to burn. Over the next several hours, the fire raced along cables like a fuse on a stick of dynamite in a cartoon, taking out not only many of the safety systems of the reactor where the fire occurred, but also those of a second reactor whose cables shared this spreading room. As the fire burned the plastic insulation coating off the cables, the raw metal wire—now exposed—could easily touch other wires, leading to electrical shorts that disabled vital safety equipment.

It took hours for plant engineers and operators to determine how best to arrest the blaze, confusion that wasted precious time and allowed more and more systems to burn.

As we all learn as children, water and live electric wires can be a dangerous combination, and so the plant operators feared that water used to douse the flames would react with the exposed wiring of the now-burned cables. Eventually they did use water, and the fire was extinguished, but not before causing significant damage to the plant's vital systems, despite the fact that the actual fire progressed only a short distance. The primary emergency cooling systems were rendered useless, forcing the plant to shut down for over a year.

The incident alerted the industry and the NRC to the fact that fires could no longer be treated as merely a company problem. They were a public safety threat. This realization led to a comprehensive rewrite of the agency's fire safety standards—standards that would then go unenforced for decades.

Nothing demonstrates the culture clash in approaches to nuclear regulation more clearly than the efforts to address the Browns Ferry fire. On one side were the traditional nuclear safety standards, "deterministic requirements" setting out lists of rules that must be followed. On the other was a new class of requirements that would be developed later relying on computer models, "risk-informed, performance-based rules."

The deterministic rules were structured exactly as you would think they should be: you define a set of do's and don'ts, laying out minimum or maximum performance expectations for reactors and other plant equipment. This makes it simple to differentiate good behavior from bad. The safety authority can easily know whether a plant is in compliance. Deterministic standards are used in everyday life; think of the speed limit on roads. They're sturdy but inflexible, and so they may lose relevance as knowledge and technology evolve.

Nuclear safety regulators determined these standards in response to a design basis accident: a limited set of accident scenarios they envisioned for what could go wrong. Designers would then develop safety systems to respond to these accidents. Because these scenarios were limited in number, early designers also tried to add in extra protections to address the limitations of the design basis accident approach.

After the Browns Ferry fire, the agency designed a straightforward approach to safeguard plants against a typical fire that could spread throughout the facility, wiping out many systems. The rules were simple, so simple that I could easily remember and recite them. As the Browns Ferry fire showed, the plant's most vulnerable elements were the power and control cables that ran throughout the building like nerves in the human body. To address this, the new deterministic rules called for separation: keeping combustibles far away from one another. That way a fire confined to one spot might disable some but not all of the safety systems in a plant.

The problem was that not all systems could be separated. Unless plants were going to be completely redesigned to isolate each independent safety system in a separate control room, all the cables for all the equipment would coalesce in one room. This meant that in addition to separating everything that could be separated, you needed a way to prevent fires from spreading in places where you could not achieve separation. So the agency added another requirement: systems that could not be sufficiently separated had to be protected against fires. Either safety systems had to be separated from one another by twenty feet, or the plant had to have each system protected by a barrier that could withstand a nearby fire for three hours, or the plant had to have systems protected by a barrier that could withstand a fire for one hour if there was also a fire suppression and detection system nearby. There was one more requirement too: there had to be an alternate control room in case the main control room was disabled. That was basically it: a solid solution, *if it had been put in place.*

In principle, twenty feet of separation between vital safety equipment seems reasonable; if one piece of equipment is fifteen feet away from another, simply move one of them another five feet. But this becomes difficult when the room the equipment is in is only fifteen feet wide. And if the room is locked in like the middle piece in a jigsaw puzzle amid other rooms inside the fortress that is a nuclear power plant, then moving walls to accommodate a greater need for separation is nearly impossible.

The rules about barriers were also complicated to follow. They should have been straightforward: just wrap parallel sets of cables in a fire barrier (which looks like the plaster cast on a broken leg). But exactly how long would these barriers prevent fires from spreading. One hour? Three hours? It was hard to know for sure.

So almost as soon as the new fire safety rules were enacted, the industry challenged them in court as unworkable—not to mention a financial burden. As the case dragged on, plants with less than ideal fire safety designs continued to be built. Finally, after years of debate, the courts eventually upheld the rules put in place after Browns Ferry, but only because the NRC promised to be flexible, allowing companies exemptions to pursue alternative approaches to preventing fires from spreading. And so the great fire regulation exemption marathon began. Over the subsequent decades, some plants would have hundreds of exemptions, many of them never even reviewed by the NRC. It was as if every citizen in a town were entitled to tell the police chief which rules to enforce. There was chaos.

Order was seemingly restored only in 2004, when the agency finalized new, modern fire safety regulations based on risk-informed, performance-based models. In this approach, scientists and engineers assign a probability to various parts of the plant failing, then they calculate different combinations of failures one by one, learning about the consequences of the failure and the likelihood of its happening. Some failures might combine to bring down the whole plant and release radiation into the environment; others might be easily corrected by the plant's safety systems. Using these calculations, not only could plant designers and safety reviewers get a better sense of what could happen, but they began to understand how often some terrible things really might happen.

Safety rules would no longer be simple and clear, like those earlier requirements for twenty feet of separation between systems or barriers that could withstand fire for three hours. In place of rules, there would now be computer programs the industry itself developed and ran.

From nearly the beginning of the nuclear power era, proponents have sought a way to mitigate the hazards of nuclear power. In their view, even the Three Mile Island accident failed to lead to any direct impact on health: the levels of radiation released were very low, and the dangerous mistakes plant operators made were neutralized by robust plant design. According to these advocates, what the general public needed was a better sense of perspective to appreciate the virtues of nuclear power without fear.

Compare, for example, the threat of nuclear disaster with other hazards, like driving a car. Surely, the nuclear power supporters argued, a public that understood they were more likely to die in a car accident than from an accident at a nuclear power plant would come to embrace nuclear technology. To spread this message, the industry turned to this new method of safety assessment: using computer models to calculate the odds of various accident scenarios. Rather than enhancing public knowledge and better communicating the hazards of nuclear power, however, these models became a way for the industry to argue that nuclear power plants were subjected to burdensome and ineffective regulations.

During Senator Pete Domenici's push to weaken the authority of the NRC in the late 1990s, he advocated for more reliance on voluntary, risk-informed, performance-based standards, shifting the responsibility for oversight from the agency to the industry. Think of the German Autobahn, where you can drive as fast as you think your car will go safely. The difference is that a car crash may harm the people involved, but in the event of a nuclear plant accident, radiation can harm many, many people, and contaminate the area's water supply and soil.

The NRC at the time agreed with Domenici. When I

joined the commission in 2005, it was still trying to encourage power plant owners to adopt these voluntary safety standards. Of course, voluntary standards would be accepted only if they worked in the industry's favor—namely, when they reduced regulation and saved money. In contrast, the new fire protection rules determined by computer modeling would *cost* money—tens of millions of dollars per plant—making them unattractive to most power plant owners. The agency still hoped to motivate owners to adopt these new standards, but we needed leverage. Which is when those decades of unreviewed exemptions to the fire protection rules came into play.

One of nuclear regulation's greatest oxymorons is "enforcement discretion," which seeks to balance a demand for compliance with the freedom of choice. For years, plants had been given discretion to protect against fires in their own preferred ways. You could call this "changing the rules in the middle of the game when you are winning to let the other team win." Whatever you call it, the NRC shied away from making the industry do something its own regulations required, opting instead to let the industry play around with alternative approaches to fire safety, alternatives that were usually less strict. But the plants always had an excuse for their transgression, and the agency was always caving.

It's worth emphasizing: these were fire safety regulations the nuclear power industry itself had developed. Why was it so difficult to convince them to support their own standards?

The battle began in November 2010, six years after the voluntary standards were adopted, with a letter sent to the NRC's senior managers by the Nuclear Energy Institute.

The industry wanted to stretch out the NRC's review of the plants' technical applications (showing the modifications they would make to comply with the new approach) over many, many years. The letter cited two primary justifications: the considerable resources needed to complete the reviews and the limitations of the computer models used to make the safety calculations.

I didn't see this letter when it first came in. By the time the plan trickled up to me, it had become known as "the watermelon chart" because the proposed schedule for reviews was shown on a complicated chart with narrow green bands and wide red ones. Needless to say, more delay was not something I wanted. But from the industry's perspective, what was the rush? The country had been living with imperfect fire safety regulations for thirty years. Waiting a little longer couldn't hurt.

The industry also argued that because the computer simulations were cutting-edge, any plant wishing to use them needed to hire the best and the brightest fire protection engineers in the country. Of course, the best and the brightest also tended to be the most expensive and the least available. So the plants' inability to recruit talented safety experts became a "justifiable" reason for delay. This was enforcement discretion run amok.

The letter also noted the amount it would cost the agency—*us*—to hire staff and develop tools to analyze the plants' proposed safety measures: "The NRC staff will receive up to 23 submittals [from the plants willing to try the new approach] with unique detailed information nearly simultaneously. This would impose a significant burden on the staff, create a flood of requests for additional information to licensees and expend licensee and NRC staff resources inefficiently." The industry, senior managers were supposed to believe, was only looking out for the NRC's best interests; the Nuclear Energy Institute was dismayed by the thought of wasting agency time and effort. As the person legally empowered to ensure the responsible expenditure of agency resources, I was taken aback by this arrogance.

Giving the industry what it wanted struck me as irresponsible. So I told the NRC staff there would be no more enforcement discretion. We were not going to change our plans simply because the utilities sent us a letter telling us to ignore our standards so that we wouldn't waste our time.

What happened over the next several weeks was more brutal than Roman imperial succession. My first task was to find out why it had taken a letter from the NEI for me to learn that the agency was ill equipped to review the applications I had been pushing the industry to deliver. It turned out that the problem lay with a few senior NRC managers who in my view had lost their desire to push back on the industry; they had become comfortable with the industry's requests to delay and defer. But I refused to accept their defeatism. I told them to come back to me with a plan to make the application process work under the schedule we had developed.

The other commissioners took the industry's side. As I was trying to gather information to remedy our lack of preparation, they were suggesting to the agency's staff that the industry should be allowed more time. Caught between the power of the chairman and the pressure of the commissioners, the agency staff was being pulled in opposite directions.

The industry now also had a powerful reason to request delay: it seemed possible that we truly wouldn't have enough staff to complete our review of the applications in a timely manner.

Jack Grobe was a senior manager at the agency when I was there. He could be brusque but was dedicated and in charge of the effort to update fire protections. He sat one day in my office in a wingback chair across from me. What he said startled me: "Mr. Chairman, I have let you down."

"What do you mean?" I asked.

"We're not going to be able to keep to the schedule for the fire safety review."

"There's no way?"

"If there's anyone who could do it, it's me. We tried everything, but it's not going to work."

That exchange told me more about the people who make up the Nuclear Regulatory Commission than any of the meetings I had held. This was no dispassionate discussion of rules and limitations. It was a personal moment, a human moment that showed the hearts of those I worked with lay where I believed they did: in support of public health and safety. The agency wanted to stand up to the industry; its staff did battle with the industry every day. With people like Jack Grobe employed by the commission, the industry could not win forever. But they had won this day.

Jack's simple honesty allowed me to accept that, despite my disappointment, the schedule for improved fire protections would have to be postponed, even if this meant the improvements might never happen. Today the agency is finally close to completing this work, albeit with the usual weakening that comes from the industry's foot-dragging.

CHAPTER 4

Nevada Roulette: Ending the Yucca Mountain Charade

Nuclear waste is almost impossible to get rid of. The current plan is to bury it and forget it, but this hasn't worked. Let me explain why.

After it sits in a reactor for up to two years, spent nuclear fuel is a poison of radioactive materials that will stay dangerous for hundreds of thousands of years. Early on, scientists and engineers determined that the best way to dispose of this material was to isolate it from people and the environment for that long. Their thinking was straightforward. Civilizations have been around for tens of thousands of years; rocks have been around for much longer. In fact some rock structures haven't changed much in thousands upon thousands of years. So why shouldn't nuclear plant operators find or build a cave surrounded by thick rocks to dispose of radioactive material? This bury-it-and-forget-it approach remains the plan today. But it's a plan that isn't being carried out. Right now, spent nuclear fuel is stored at the reactor sites that produced it. And although the nuclear industry objects to this as a permanent solution, with proper monitoring the fuel can be safely stored on-site for centuries at least.

Still, the industry wanted someplace to store spent fuel

where it could be out of sight and out of mind. The most tantalizing solution came in 1987 with the Yucca Mountain project, which was conceived as the largest government-civilian public works effort in the history of the United States. Yet like a mirage in the Las Vegas desert, the solution was never as solid as its proponents projected. Its greatest virtue was simply that it was *a* solution. When nuclear power plant operators were asked, "What are you going to do with all that dangerous spent nuclear fuel?," they could reply, "That's simple. We're going to ship it to Yucca Mountain."

Yucca Mountain is actually a ridge of mountains that were formed by volcanoes millions of years ago. The ridge sits on the edge of an area in the Nevada desert that was used to test nuclear weapons, known in the 1980s as the Nevada Test Site. About the size of Rhode Island, the site is surrounded by the even larger Nellis Air Force Range, which is used for simulated air force battles and weapons tests. On the face of it, this seems like a perfect place to put a nuclear fuel disposal site: in the middle of the desert, surrounded by military testing and training areas. But the area is just over an hour's drive from Las Vegas, one of the fastest growing metropolitan areas in the United States over the past decade and a half.

Because disposing of radioactive material seemed too difficult for the private sector in the early years of the nuclear power industry, the federal government stepped in to take responsibility. Congress passed the Nuclear Waste Policy Act of 1982, creating an apolitical framework to identify, license, and build facilities to permanently dispose of used nuclear fuel. The law was clear: the Department of Energy would identify two sites to store nuclear waste, one in the eastern half of the United States, the other in the western half. Given that most nuclear power plants were situated in the eastern half of the United States, there really was no reason to look to the west, except for the fact that the wide-open spaces seemed better suited to isolating dangerous material.

Soon the search for sites in the east and west turned into a search for sites in the west alone. Technical standards used to identify the geology best suited to storing waste were set aside in the face of pressure to identify any site at all. Impatient to address the waste disposal question in the aftermath of the Three Mile Island accident, Congress voted to select Yucca Mountain as the only site for further consideration, scrapping the earlier objective, technical approach. The decision was simple: one state had to host the repository, and many states had to get rid of their waste. In the 1980s Nevada had a weak congressional delegation and a large swath of federal land. It was a political decision, and the state's residents and elected officials knew it. The dice were loaded to ensure Nevada always rolled snake eyes.

Congress chose Yucca Mountain as the disposal site before I graduated from high school, so I had never even heard the words "Yucca Mountain" until I started working for Congressman Markey, a longtime critic of the site. But after spending a decade and a half working on the final act in the Yucca Mountain story, I was so involved in the project that my tombstone will likely read "The guy who killed the Yucca Mountain project." Over that epitaph will be spray-painted graffiti: "Clown. Bozo. Inmate." Those are just some of the epithets hurled at me during those years.

When I started working with Senator Reid, he asked me

to find a way to stop the project. My efforts on his behalf took place early in President George W. Bush's first term in office. During the presidential campaign, Bush had declared that any decision to move forward on Yucca Mountain would have to be based on "sound science," a phrase that Nevadans heard as code for opposition to the project. So when Bush decided to move forward with plans to construct the repository in 2002, Reid thought the president had lied—and he wanted to tell people so. I worked with Tessa Hafen, Reid's congenial and effective press secretary, to prep him for an interview with *60 Minutes*, urging him to avoid using the word "lied." *Misled. Confused. Misspoke.* Anything but "lied" would have been preferable. Making so strong a charge against the president would not help Reid's standing with Republicans in Nevada. But when the question came up, he seemed to glance at us standing off camera. Then he smiled mischievously and said, "He lied." The clip never made it into the program, but it made me appreciate the bluntness and fearlessness that drove Senator Reid.

It was no accident that six years later, Senator Obama opposed the Yucca Mountain project as a presidential candidate. As a senator from Illinois, he would have had good reason to support the project; no state has more nuclear reactors than his. But before the 2008 election, Reid orchestrated moving up the Nevada caucus date so that it became the second major Democratic nominating event. That put Nevada's issues front and center early in the campaign. Every presidential candidate had to learn two things. The first was that the state's name is pronounced Nev-AD-a, not Nev-AH-da. The second was to say Yuck-a Mountain, not Yook-a Mountain, and as a corollary to say, "I oppose

Yuck-A Mountain." That's just what Obama said during a precaucus debate in November 2007.

There were many technical, political, and safety reasons why the site was not ideal for nuclear waste disposal. In fact Yucca Mountain failed to meet the original geological criteria. The rock that would hold the nuclear waste allowed far too much water to penetrate; water would eventually free the radiation and carry it elsewhere. In addition, safety studies that showed the site to be acceptable were based on infeasible computer simulations projecting radiation hazards over millions of years. Realistically forecasting the complex, long-term behavior of spent nuclear fuel in underground facilities is scientifically impossible.

After a lengthy primary campaign and a historic election, the new president was quickly confronted with how to make good on his promise, as Yucca Mountain was already under licensing review at the Nuclear Regulatory Commission. The Department of Energy, the owner and developer of the Yucca Mountain storage facility, had sent a license application to the NRC in June 2008, just six months before the end of President Bush's second term.

While Obama's promise to oppose the facility was popular in Nevada, his position was extremely unpopular with the nuclear power industry and his own secretary of energy, Steven Chu, who seemed uncomfortable stopping the project. Chu was a strong supporter of nuclear power and wanted to find a disposal site for its spent fuel. In the months before I became chairman of the NRC, I would hear often from Senator Reid's staff that Secretary Chu was concerned about the ramifications of withdrawing the license application submitted under the Bush administration. As with almost all nuclear power matters, nothing was ever simple, especially if the industry got involved.

In the fall of 2009, by which time I had been chairman for almost six months, the president nominated Bill Magwood to serve on the commission. Magwood, a former Department of Energy official once in charge of the nuclear energy development program, had recently given an interview in which he downplayed the problems at Yucca Mountain and appeared to indicate his support for the project. This statement put Reid in an uncomfortable position, because Magwood's nomination appeared to undermine the president's commitment to oppose Yucca Mountain. Things were particularly tricky because, as the Senate majority leader, Reid played a large role in getting new commissioners confirmed. Opposing a nominee selected by a president from his own party could lead to attacks in the Nevada press. Fortunately, the issue was neatly resolved: Magwood would be appointed to the commission, and the Department of Energy would pull the Yucca Mountain license application back from the NRC and add it to a warehouse full of failed public policy projects.

The Yucca Mountain debate shows how what constitutes "safety" is often determined by political, not just scientific, judgments. And yet the concerns of people outside the industry are regularly dismissed and ridiculed on the basis of elusive numbers and calculations.

When we think of safety limits, we often think of precise numbers that tell us how much of something we can be exposed to without harm: Don't eat more than so many ounces of swordfish a week, we're told, or you risk exposure to toxic levels of mercury. These hard numbers and concise statements give the impression that these figures are based on strictly scientific calculations. In truth, safety limits never are.

The reason is quite simple: safety is not an objective statement of scientific truth. Safety is a subjective determination made by societies—or their designated representatives—about the acceptable behaviors that companies and individuals can engage in. There is no textbook that says how much radiation exposure is too much. There certainly are textbooks that set forth the various ways radiation can cause harm, but societies themselves must decide at what point they want to control harm. We allow people to smoke, although tobacco is a known carcinogen, because as a society we have decided that use of this hazard is best left to individual choice. We allow unlimited alcohol consumption for people over the age of twenty-one, even though there is no textbook that puts forth twenty-one as the appropriate age for responsible drinking. In fact, this limit only came about in 1984. Some countries do not even have a set legal drinking age.

Because safety, especially as it relates to public health, is often informed by medical and scientific data, we tend to believe that safety standards are also determined by science. But societal norms, traditions, customs, and politics play a role in establishing those standards. Safety decisions are public policy decisions that endeavor to balance the interests of all the competing elements in a society.

The decisions we make regarding nuclear facilities are no different. There is no scientific reason that the NRC sets the safe level of radiation exposure at 100 millirem (mrem).

Yes, this is a policy decision informed by science, but other factors also come into play. That the limit is a nice round number like 100mrem should be a clue to its arbitrariness. It would be an amazing feat if physics and biology had conspired in such a way that safe radiation exposure could be expressed so simply using the number 100 and not, say, 102.36493. It is an unwritten rule of safety policymaking that, whenever possible, you work with simple numerical expressions. This is not scientific. This is practical.

In the United States, federal government agencies are required by law to offer rules for public consideration and review during an official comment period. The very nature of this process leads to a sorting of scientific, monetary, practical, and political factors. The Yucca Mountain project was no exception.

Along with Bill Magwood, the president also nominated Bill Ostendorff, a Republican, and George Apostolakis, a Democrat, to fill vacancies on the commission in the fall of 2009. Once a deal with Congress was reached to confirm all three in the spring of 2010, the administration had to fulfill its part of the deal. That meant the Department of Energy had to withdraw the Yucca Mountain disposal site's license application from the NRC. Since the application had already been granted an official hearing by the NRC, the request to withdraw had to be made through a formal, judicial-like process.

At the center of this story is one of the many boards that does the initial work of reviewing the agency's formal licensing decisions. These boards are made up of judges with legal and technical backgrounds who are responsible for conducting the complex proceedings related to the licensing of nuclear facilities. Their decisions are subject to review by the full five-person commission, which acts like a court of appeals. Absent action by the commission, the decisions of these judges stand, although almost all significant decisions are appealed to and reviewed by the full commission.

The NRC judges overseeing the license review complicated the Department of Energy's request to withdraw the Yucca Mountain application. Deciding they did not know what to do with the request, they bypassed the commission and asked a federal court to decide. This was unprecedented. It fell to me to mobilize the commission to tell the board to go back and make a decision on the withdrawal request, one way or the other. Before doing so, I asked a few attorneys whether the licensing board could legally deny DOE's request. All told me no. Ideally, I would have preferred for the commission to simply address the question itself, but I met opposition from the other commissioners on this point. I would realize later that I had missed a red flag about their true feelings: my fellow commissioners wanted Yucca to proceed.

The licensing board went back to work and, to my surprise, rejected the DOE's request. Despite the order to close the Yucca Mountain offices, shutter the facility, and release the associated staff, the application to review the license would go forward. This was absurd—like a fire marshal telling developers of a shopping mall that they could not cancel construction because of pending inspections.

The NRC was supposed to be an impartial arbiter of the safety of the country's nuclear facilities. When the agency forced an applicant to defend an application that the applicant no longer believed in, the agency became invested in the outcome. This shredded the agency's impartiality. More important, this decision—and the collective cheer it elicited from the supporters of the Yucca Mountain project—further reinforced my belief that supporters of nuclear power viewed the agency as a tool to promote the nuclear industry rather than a force to regulate it. Yucca Mountain was, after all, essential to the industry's success. Without a permanent depository for used nuclear fuel, it would continue to face challenges to its efforts to operate and possibly even expand.

None of the licensing board's shenanigans, however, seemed to matter to the Department of Energy. In February 2010, while the board was still deliberating, the DOE closed down the Yucca Mountain site. Thousands of contractors and federal workers were terminated. The project was over. This action only made the licensing board's decision more nonsensical. How could the DOE defend an application for a project that no longer existed? But the industry was desperate to revive the project, and they had a commission willing to try—even if that led to a showdown between the president and the industry.

When the secretary of energy closed Yucca Mountain, he applied the funds for that project to other programs. But to halt the Yucca Mountain project for good, Congress would need to stop providing money for it. The ultimate arbiter of any federal project is Congress, because Congress funds the government.

And Congress had been doing a pretty bad job lately of performing that basic responsibility. In years past it would pass about a dozen pieces of legislation to fund all parts of the federal government, each bill covering the same agencies year after year. More recently, as the process began to break down, these bills became lumped into one or two massive spending bills that covered the entire federal government, making it difficult for individual agencies to know what their annual financial resources would be. The situation was even worse in 2010. Congress failed to pass its omnibus bill before the end of the fiscal year that October. Instead it passed something called a continuing resolution, providing short-term funding at levels consistent with the previous year.

As the DOE and the NRC's licensing boards were wrangling over what to do with the Yucca Mountain project, I was faced with a difficult decision about agency expenditures. Based on the previous year's funding, the agency had money for the Yucca Mountain license review. (Even though the judges had ruled to prevent the withdrawal of the DOE's application, the commission still had to review the licensing board's decision.) That funding, however, would likely be revoked once new spending bills were finally passed. If we went ahead with the review, the agency would probably spend money we would never receive. I ended up choosing the more conservative approach: the agency would stop work on its review, just as the Department of Energy had stopped work on the project. Any further work by the agency's staff would be unnecessary.

Here is where the story gets interesting. At most independent regulatory commissions, the chairman has the executive authority to make these kinds of decisions. Even at the NRC, the chairman has this authority—except when the commissioners make an argument that the chairman doesn't.

This strange distribution of authority came about after the Three Mile Island accident, which revealed the agency's lack of any clear decision-making authority during an emergency. President Carter proposed strengthening the chairman's role as the agency's CEO, with the commission setting policy and the chairman executing that policy. Congress, however, balked. It accepted the president's proposal with a caveat: in cases where executive action might be reasonably construed to be policy, the commissioners could vote to determine whether the chairman's action was policy or execution. Over time this provision came to mean that any action by the chairman was subject to a vote. This dysfunction would come to play an important role during the response to the Fukushima accident. But for now, before I halted the agency's review of the Yucca Mountain license application, I wanted to make sure a majority of my colleagues would support me. So I asked them.

I began with George Apostolakis. As a former nuclear engineering professor and technical advisor to the agency, he was inexperienced in nuclear industry politicking. So I talked him through the issue. I told him I was going to stop work on the staff review of the Yucca Mountain project because it was about to be eliminated from the budget and asked if he would support that decision. He told me he would.

Next I turned to Bill Magwood, who, although appointed by President Obama, had often been an antagonist during the short time he had served with me on the commission. He took a few days to consider my decision but agreed to support me.

After receiving Magwood's commitment, I had the

votes I needed to move forward to close down the staff work. But I still wanted to talk to the other two commissioners, Bill Ostendorff and Kristine Svinicki. Ostendorff is a former navy submarine captain, used to being in charge of a crew and unaccustomed to the more collaborative role of commissioner. When I called to tell him my plan, he was so loudly irate I had to hold the phone several inches from my ear. I told him that I would think a bit more about his opposition before making a final decision; this was a mistake, because it gave Ostendorff and Svinicki time to pressure Magwood and Apostolakis to reverse their positions. And so the conflict began.

As soon as I made the decision to move forward, the Democratic commissioners buckled, having been pressured by their Republican colleagues. Apostolakis decided that he had a previously unknown conflict of interest from earlier work he had done that was vaguely related to the project and so could not vote on any matters related to Yucca Mountain. That got him out of the cross fire. The vote to end the licensing review now rested with Magwood, whom I convinced not to walk away from his commitment. And so Magwood came up with a novel way to support me without looking like he was supporting me: he recused himself from the vote. With Apostolakis and Magwood unwilling to participate, there were not enough commissioners for an official vote, and so no action could be taken to reverse my decision.

This whole episode was particularly surprising because the fate of Yucca Mountain was never really in doubt. Once the president, through the secretary of energy, had shut down the Yucca Mountain program, once Congress passed its spending bills it would receive no further funding. It was a done deal. There was no need to review the application.

Shortly after my decision to shut down the agency review, Congress passed a final budget that provided no funding for the review. After thirty-five years, the Yucca Mountain project was over. Now it was simply a matter of wrapping up the work and conducting the difficult task of reassigning staff.

But the matter would not die. Not even Fukushima, the world's third major nuclear power accident, would put a stop to the obsession some members of Congress had with reviving the Yucca Mountain project. Sitting in the House Energy and Commerce Committee hearing room in March 2011 to testify about Japan just days after the start of the accident, I was questioned by John Shimkus of Illinois about my termination of the agency's review of Yucca Mountain.

The Yucca Mountain experience damaged my relationships with my colleagues in ways that I never anticipated. Instead of accepting the outcome as part of the rough-and-tumble world of nuclear policy, the commissioners claimed I had abused my authority. They engaged the agency's inspector general to investigate me and my decision.

Over the coming months, the inspector general would spend considerable resources digging into this decision, which puzzled me because the decisions of the commission are supposed to be beyond his purview. Whether I acted legally or illegally was a determination to be made by the agency's general counsel, who had signed off on my decision. If there was any issue to be investigated, I felt, it was how the industry had tried so aggressively to influence the commission. Its representatives had used their most powerful tool—sympathetic senators and representatives—to convince me to allow the hearing to go forward, despite the Department of Energy's decision to shut the project down. Their congressional allies pressured and even threatened me in meetings, letters, and public hearings. I was accused of breaking the law, of malfeasance, and of downright badness.

After Congress finally passed the appropriation bills without any new funds for the NRC's review, it became obvious that it was time to move on, and the commission voted to wrap up the remaining licensing hearing. I was ultimately exonerated by the inspector general, which only agitated the other commissioners more.

The griping and politicking about Yucca Mountain in the summer of 2011, even after Congress defunded the program, reveal the extraordinary power of the nuclear industry. No other industry is able to complain so loudly that someone else has failed to take care of its waste. Nuclear industry proponents described the closing of the project as "political," but when they needed federal support to build new reactors, they called the federal government's actions "policy-driven." Both are just the decisions of power brokers striving to achieve the outcome they want.

You would think that it would make sense to suspend nuclear power projects until we know what to do with the waste they create. But that ignores the unwritten laws of nuclear politics: Regulators are there to ensure safety, but they must always balance that need against the imperative that the industry survive.

As waste piles up, we leave behind dangerous materials that later generations will eventually have to confront. The short-term solution—leaving it where it is—can certainly be accomplished with minimal hazard to the public. But such solutions require active maintenance and monitoring by a less than willing industry. This is already an organizational and financial burden. In thirty thousand years, when these companies no longer exist, who will be responsible for this material?

There is only one logical answer: we must stop generating nuclear waste, and that means we must stop using nuclear power. I wish that as chairman I'd had the courage to say this, but my courage had its limits. I knew the backlash that would come if the chairman of the NRC were to admit our country should stop producing nuclear power.

My failure to continue the fight after shutting down Yucca Mountain left me wanting never to regret another decision I made. In 2012 a debate over the licensing of new reactors would give me a chance to redeem my conscience.

CHAPTER 5

Accidents Do Happen: The Tragedy of Fukushima

Deep under the ocean off the coast of Japan, two of the plates that make up the surface of the Earth were grinding and pressing against each other like sumo wrestlers locked in a tussle. Unknown to the operators at the Fukushima Daiichi nuclear power plant one hundred miles away, one of these plates let go of its grip. As one piece of the Earth's crust slid over the top of the other, seismic monitors at stations across the globe came to life, their electronic needles bouncing up and down, tracing out the pattern of a gigantic, magnitude 9 earthquake. It was the largest ever recorded in Japan, and one of the largest ever recorded in the world. Yet it was not the worst problem about to crash down on the Fukushima Daiichi nuclear station.

The upheaval of the ocean floor that caused the earthquake also sent walls of water nearly fifty feet high hurtling toward the beach outside the Fukushima plant. The tsunami lifted large boats and carried them miles inland. It swept away homes and businesses and people, leaving behind nothing but fields. Tens of thousands were killed.

Despite the tsunami's thunderous arrival, the nuclear accident would progress slowly and silently over the days after the waves battered the fortified shell around the reactors at the Fukushima Daiichi plant. The nuclear reactors would be disabled from the inside like a body affected by a small but swiftly growing tumor. By the time disaster had run its course, over one hundred thousand people had been forced from their homes. Many of them will never be able to return. The Japanese economy collapsed, and the government eventually halted operations at all of the nearly sixty nuclear reactors in Japan.

The accident spurred three independent investigations, all of which came to a similar conclusion: the nuclear power regulators were too accommodating to those they were supposed to regulate. They worked together to create what one report called "a nuclear village," not an idyllic hamlet where business and government worked in harmony for the good of all but a corrupt, toxic environment.

Six white boxes, each about two hundred feet high, housed the six nuclear reactors at the Fukushima Daiichi site. Four of them were near each other, nestled up close to the shore, standing watch over the water to their east. The other two were more than a football field's distance away. Behind the plant were sloping hills that would trap the waters and the force of the tsunami. Before the accident, Tokyo, one of the most heavily populated cities in the world, at 150 miles' distance seemed far away. As the accident progressed, the city became uncomfortably close.

Originally designed by General Electric, Toshiba, and Hitachi in the 1960s, the Fukushima reactors looked modern and serene, their white-and-baby-blue color scheme helping them blend into the horizon without spoiling the ocean view. Like most nuclear power plants the world over, large metal transmission towers supplied electricity to make the plant's production and safety equipment run. The earthquake knocked these towers down.

But reactors are supposed to be prepared for this scenario. Most nuclear power plants—including the reactors at Fukushima—house large diesel fuel–powered engines, like something you would see on a freight train or large merchant ship. Powerful and intimidating, these engines would surely provide all the electricity needed in an emergency. Only, in a design failure later attributed to coziness between government regulators and the powerful utilities operating the plants, key parts of the diesel engines were installed in the tsunami's path, rendering them useless in the crisis.

Inside the plant, operators were unaware of the danger that would come from the tsunami. The earthquake, however, they could feel. Following the shaking and rolling, the operators began emergency safety procedures to stop the reactor engine and start the cool-down process. The most important action occurred automatically: the plant instantly stopped the nuclear fission reaction responsible for creating electricity.

The computers monitoring the power supply coming into the plant detected a drop in electricity, and the diesel engines that supply emergency power came to life with a slow rumble, like an elephant climbing to its feet. The basic systems were working. The operators began to follow rigorous emergency procedures. They rifled through checklists and ticked off activities that would surely settle the reactors into a safe configuration. Nuclear power plants, after all, were no longer supposed to have accidents. Certainly a response to a common occurrence like an earthquake (Japan is one of the most seismically active countries in the world) was within the plant's defensive capabilities. But this earthquake was different because of the tsunami it created.

The largest wave came rushing toward the reactor site less than an hour later, slamming into the buildings housing the reactors and the low sea wall that buffered the plant against normal ocean waves. The hills kept the water, more than thirteen feet deep by the last wave, corralled around the reactors. The force of the waves pushed water into the plant, exposing vital electronic equipment to salt water. All but one of the diesel engines were destroyed, leaving the station almost completely without electricity. With no light to see by and with their instruments disabled, the operators were no longer in control of the reactors.

Part of my job as the NRC chairman was to respond when an emergency arose at a nuclear power plant anywhere in the world; I was always on call. But to spare me constant interruptions, the gurus who staffed our operations center—referred to as HOOs and HEROs, acronyms for headquarters operations officers and headquarters emergency response officers—had a high threshold for calling me at home, especially in the early morning. So when my phone rang early on Friday, March 11, 2011, while I was still in bed, I knew the call was serious.

The information relayed to me was simple. There had been an extremely large earthquake in Japan, and two nuclear power plants on the West Coast of America were on alert for a tsunami.

Tsunami waves can travel thousands of miles. With little information about the size and strength of the one that was surely to come, our focus was on two nuclear power plants sitting by the Pacific Ocean in California. As the story unfolded in Japan, however, we would learn once again that no emergency is predictable.

In the first few hours, the hints of impending calamity were not easy to spot. There was certainly a sense that something was happening in Japan, but the details were fuzzy. So I set about my day with the coming arrival of the tsunami on America's West Coast tickling the back of my mind. A briefing at 10:00 resolved my concerns: the tsunami's impact on the California plants was minimal. Now my attention could shift to the plight of the power plants in Japan. But I wasn't too worried. Even if there were problems, I expected the Japanese government to be able to respond. After all, nuclear power plant accidents no longer happened.

By noon, however, the magnitude of the disaster was becoming evident. The U.S. military, which has a large presence in Japan, had been deployed to help find people amid the wreckage. At the end of the day, instead of leaving for home, I participated in the first of many National Security Council teleconferences with the many U.S. departments and agencies assisting in the response.

By late Friday evening—early Saturday morning in Japan—it was clear that the safety systems at several reactor sites were either disabled or vulnerable to failure. Three of the plants initially believed to be threatened were restored to a safe configuration. (One of those sites, Fukushima Daiini, was only a few miles from Fukushima Daiichi.) But four reactors at a fourth site were affected.

As I had moved from my office to the futuristic cock-
pit of our emergency response center earlier in the day,
I'd expected to receive an update on the reactors and then
head home after a long week. Instead I sat at the head of the
conference table for hours as the dire situation in Japan became clearer. In between the reports of escalating alarm, I
reconsidered much of what I thought I knew about nuclear
power plants.

In these early days after the earthquake, the biggest challenge I faced was a lack of reliable information. As studies
would later show, many of the instruments intended to
report the reactors' vital signs were damaged, disabled, or
otherwise unusable. Our analysts were left to improvise.
As reports that the plant now lacked basic safety protection
trickled out, they tuned in to news reports, reached out to
colleagues in Japan, and contacted industry experts in the
United States to fill in the gaps in their understanding. As
the accident progressed, some journalists, policymakers,
and activists began to wonder if the tight flow of information was the result of government secrecy intended to min-
imize the extent of the accident. But the real culprit was the
lack of information.

To get the information we required to assess the situation at Fukushima and the threat to the United States, we
needed people closer to the accident; someone had to go
to Japan as quickly as possible. The federal government was
mobilizing transportation for officials from many parts of
the government, from search-and-rescue responders to full
naval fleets. When we were offered a seat on the plane a lone

NRC staff member rushed from our emergency response center to his home to get his passport, then to Dulles International Airport. He would be the first of what would become a vital team of about a dozen nuclear safety experts who would spend months in Japan. We had prepared multiple accident scenarios in the emergency response center, but we had never rehearsed reacting to a disaster at the plant of a political ally ten thousand miles away.

As the accident began its terrible unfolding, my primary concerns were communicating the situation and preparing to address its impact for the United States. The full extent of the damage and the potential for hazardous radiation to be spewed into the air were not known. All my training and experience, bolstered by the knowledge and expertise of the agency's staff, suggested the accident would end soon with limited public health consequences. But we had to prepare for the worst.

Throughout that first weekend, I told myself to remember two important principles. First, the people closest to the situation had the best access to information. In this case, that was the Japanese government; our own communications should follow their lead. Second, my role as a regulator was to provide support, guidance, and direction—in that order. I believed that the NRC should rarely move beyond support, which is to say making government resources, expertise, and equipment available to plants in trouble. This is because ultimately the owners of nuclear plants, which in the United States and Japan are private companies, are the ones responsible for dealing with accidents. These companies often make significant profits, and with this success comes the responsibility for dealing with any accidents that may occur. As a result of these two principles, we said very little during the first weekend, despite tremendous pressure from nuclear industry proponents who wanted us to speak out. The message they wished us to deliver was that everything was fine. I was not convinced it was.

During those first few days, we were all complacent after years without a major accident in the U.S. Even the exercises conducted in the room from which I was leading the agency's response to Fukushima had exacerbated this. The agency's creative engineers had designed complicated accident scenarios involving severely degraded reactors, but not ones as degraded as the Fukushima reactors now were. The designers would have to manipulate and hyper-extend the calamities to get the reactors to a point where they would emit radiation into the environment. It was as if the designers were animal trainers trying to get a tranquilized elephant to its feet. Even then, the damaged reactors would release some small amount of radiation and then be restored to normal, neatly terminating the accident. This left the agency conditioned to think that truly bad events were impossible in modern industrialized countries like the United States and Japan.

As I mentioned, in Fukushima the biggest problem came from the loss of power. Ironically, many nuclear power facilities do not use the electricity they produce to power their own safety systems. So if the power coming into the site is lost, the plant must rely on backup power systems. Loss of off-site power was frequently shown in safety analyses to be one of the worst accident scenarios. That scenario was happening now.

Nuclear power plants are prepared for power loss; they

have massive diesel engines on-site. Every plant in the United States has at least two, each one about the size of a small locomotive engine. (Some come originally from diesel-powered ships.) To run, they need a few simple things: some form of cooling so they don't overheat and fail (a common requirement for almost any major piece of machinery), diesel fuel, and a system to distribute electricity from the diesel generator to the rest of the plant.

In Japan nearly all aspects of the power supply system were damaged. The earthquake toppled some of the transmission lines that supplied outside power to the reactor, requiring the use of the on-site diesel generators. The tsunami destroyed the fuel supplies for the diesel generators, leaving them to function with a limited supply that was quickly exhausted. The tsunami also flooded a large area of the plant, disabling the electrical distribution system. One of these failures by itself would have caused significant damage to the reactors. Together they formed an almost insurmountable obstacle to recovery.

At the Rockville operations center on Friday, March 11, we were not fully aware of the extent of the damage. Communication between the agency and those in Japan was limited. Our emergency response center was calm, and the silence extended far beyond that room, all the way to the U.S. Embassy in Japan.

To keep focused during the periods of waiting, I asked the agency staff to consider a number of scenarios to determine the potential impacts should the accident worsen. There was, however, very little to suggest this would happen. After all, they simply had to get some water into the reactor core and some electricity flowing to the pumps to move the water around. In the minds of most nuclear professionals, this was not that hard.

Then hydrogen explosions rocked the plant and startled the world.

Before the Fukushima nuclear accident, the most iconic image of this sort of explosion was the fireball that destroyed the *Hindenburg* airship in 1937. The black-and-white photographs of the massive airship floating nearly vertical, one end ablaze, symbolized the disastrous result of hydrogen mixing rapidly with oxygen. On March 12, 2011, this image of hydrogen's destructive power was joined in the public imagination by the mass of dust and debris thrown into the air from an explosion at unit 1 of the Fukushima Daiichi plant. The debris rose so high it dwarfed the reactor buildings and dramatically changed the course of the accident.

The explosion initially puzzled the agency staff—as would many occurrences over the next few weeks—because most nuclear safety professionals believed plants were effectively designed to prevent the events we were now seeing. In the United States, the NRC had dealt with hydrogen control decades earlier. The first and simplest way to thwart a blast is to remove oxygen from the scene; without it, hydrogen has no explosive dance partner. Immediately after the initial Fukushima explosions, there were no indications that the primary containment unit, into which nitrogen was injected to keep oxygen out, had leaked, allowing atmospheric oxygen to mix with the liberated hydrogen. Because of this, we at the NRC thought the explosion had to be caused by something other than hydrogen. When the staff finally accepted that the primary containment unit had been breached, the implications were severe: unit 1 was beyond salvation. We were witnessing the first major reactor accident in decades.

Vivid images that resembled a Hollywood disaster movie were circulating, and soon the press, the environmental community, and Congress would ask what the Fukushima accident meant for the United States. I knew the nuclear industry would have an answer ready: "This is a Japanese problem. American plants are safe."

To be able to respond to this inevitable question, I asked the agency staff to estimate the accident's impact on Hawaii and Alaska, the states closest to Japan, as well as American territories in the Pacific Ocean. There were very few times I anticipated any harm to America. The distances were simply too large and the amount of radioactive material expelled even in the worst cases too small for harmful amounts to reach any part of the United States. American forces and other citizens living in Japan were, of course, much closer to the source of the radiation. I felt strongly that the Japanese government could provide the best advice on how they should respond to the accident.

Because we had such limited information, the staff had to make assumptions about the condition of the reactors, the amount of radiation already being released, and the amount of radiation likely to be released overall. This figure, the source term, was limited by two divergent extremes. On one end of the spectrum, the radiation release could be zero. Absolutely none. But we knew the situation had already deteriorated beyond that point. The other extreme was for all the radioactive material in the reactor core to be released unfiltered into the environment should the containment structure fail. This was, in theory, the largest release of radiation that could occur. In reality, however, even in the worst accident, some radioactive material would flow to the basement of the reactor or stick to the walls or the pipes or the water. Still, this scenario provided a useful gauge of the most extreme contamination that could occur.

Our early findings showed the contamination could spread beyond the distance for which our computer simulation could produce results: fifty miles. This value would play an important role in the recommendations the U.S. government would issue before long.

While the staff began to formulate these assessments, the situation in Japan continued to worsen. On Monday, March 14, a second hydrogen explosion occurred in another reactor at the site.

The accident, which had now lasted more than three days, was continuing for longer than most safety experts had expected. The simulations and scenarios we had tested never envisioned an accident going on so long. Experts believed that a modern, industrialized nation would have arrested the damage by this point. Unfortunately, the damage was already irreversible. The world was now witnessing the largely unstoppable degradation of the plants.

After constant pressure to make public statements during the first days of the accident, I was pleased to hear

Monday morning that the Obama administration would address the issue during the daily briefing held by the president's press secretary, Jay Carney. I was, however, concerned that no one from the agency had been asked to attend. I raised this in my next conversation with the president's lead staffer for the issue, John Brennan, a seasoned, unflappable government veteran who had worked for most of his career at the Central Intelligence Agency, an organization he would eventually lead.

"I believe it's important for a nuclear safety regulator to be at the press briefing," I told Brennan over the phone. "It will send the message that nuclear safety is the primary concern."

"We agree," Brennan replied, "but we're worried that you represent an independent agency. We don't want to give the impression that we're influencing your decisions inappropriately."

"It would be an important symbolic statement that the administration values our input," I insisted.

Brennan said he would check and get back to me.

Knowing the drive could take at least an hour, I told him I'd hop in the car now and head down to the White House. If he didn't want me there, I'd just turn around. Before I could make it out of my office door, Brennan called back to say "Please come."

On the way to the White House I thought about the first public statement I would make about the accident. It was one thing to discuss the situation with other government officials and another to speak to the White House press corps, potentially even to the whole world.

Upon arriving at the White House, I had a lengthy discussion with Jay Carney and Dan Poneman, deputy secretary at the Department of Energy, about what each of us would say. We then entered the White House press briefing room.

I had several basic messages to deliver, honed through extensive NRC staff briefings over the previous days as I learned everything I could about the accident and its implications for plants in the U.S. Then I stood behind the White House podium with Poneman and Carney and relayed these messages to the press. First and foremost, I asserted that the Japanese government was in the best position to instruct Americans on how to act because it should have the most accurate, up-to-date information. Over the weekend, our contacts there had provided insights about the evolving status of the reactors, although information was limited as the reactors were not producing much data due to the severity of the damage. Second, based on my understanding of the nature of the accident, I declared that even in a worst-case scenario, radiation from Japan could not be concentrated enough to pose a hazard to Americans outside Japan, not even to those living in Alaska and Hawaii, after traveling thousands of miles across the ocean. Third, I offered assurances that American plants were different from those in Japan.

Those assurances to the media and TV viewers watching were, it turned out, platitudes. After several months, as I struggled to push through reforms based on the lessons of Fukushima, I would come to realize that American plants were not so different after all. It would take only a matter of hours before I began to doubt my first point as well. By Monday evening, just hours after I told the world to listen to the Japanese government, I told the U.S. government to trust our agency's experts instead.

The NRC operations center continued to monitor Japanese media and other sources, hoping for insights about the conditions of the reactors. Disturbing images and information continued to flow in, especially concerning yet another reactor at the Fukushima site. While the first unit was irreversibly damaged within just a few hours of the tsunami, two of the other reactors appeared to be harmed but controllable. Then blocky images of a fire appeared on the operation center screens. During the day on the Monday of the press conference, the world had seen more hydrogen explosions. Now, at nighttime in Japan, we watched a new fire burning in unit 4.

The NRC staff recognized this fire could be a sign of something more ominous, although they could not immediately identify what that would be. Some of our experts speculated that yet another hydrogen explosion had occurred and that the fire we were seeing was a result. Firefighters were dispatched, and the fire was eventually extinguished. We later learned that it had been external to the building containing sensitive safety systems and would therefore not further degrade the safety of the plants. But it was another challenge to workers trying to arrest the uncontrolled heating of the reactor fuel.

Another major setback would occur in unit 2. Each of the reactors at the Fukushima site is housed in a protective shell, called a drywell, made of thick steel. Viewed from the side, this structure looks like an inverted lightbulb with a halo. It is responsible for adsorbing the pressure that comes from steam produced by the intense heat of the reactor.

Imagine sealing a tea kettle once the water starts boiling. To reduce the pressure, the reactor directs the steam into a doughnut-shaped system of pipes near the bottom of the drywell, large enough for a person to climb around in. The pipes cool the steam, which condenses back into liquid water. If this system fails, you would know it right away: the pressure would drop abruptly, signaling some breach of the containment, like a giant radioactive balloon popping.

On Monday night, Bill Ruland, a wise NRC safety expert, approached me in a back room of the operations center, found a chair, and slumped into it. With graying hair shaved close and glasses that always seemed to be on the verge of falling from his face, he was just the kind of expert I wanted explaining the situation to me.

Bill's words were ominous. We had received a report indicating a loud sound coming from unit 2. Other indications showed a drop in pressure in that building.

Reactor fuel can melt only at temperatures much hotter than those needed to melt steel. Like a warm spoon scooping ice cream, this fuel could melt through the reactor vessel. If it was hot enough and there was enough of it, it could then melt through the drywell. What were the symptoms of such an event? A loud bang followed by a drop in pressure. The reactor fuel in unit 2 had likely started to melt.

Shaking his head, Bill said, "This was not supposed to happen."

So much of what was happening was not supposed to happen.

Now this second reactor would be uncontrollable. Even more radiation would be released into the environment, harming the public and hampering the emergency response. Our computer simulations were showing that land contamination could spread well beyond ten miles from the plant—an almost sacrosanct limit in the U.S. nuclear safety system. Every nuclear power plant in the U.S. had to develop emergency action plans for communities within a ten mile radius. The Japanese were now evacuating people outside that distance, although not as far out as our models were projecting the radiation could spread: as much as fifty miles in the direction of the prevailing winds. As I would learn when I traveled to Tokyo two weeks later, the Japanese were beginning to worry the contamination could spread farther still—perhaps even 150 miles, far enough to threaten Tokyo, a city of 13 million of people.

I first learned how far the radiation might spread late Monday night, during a conference call with senior government officials from the president's staff and the American ambassador to Japan. A few agency staff members and I had retreated to a room just off the main response center to talk without background distractions. While I was speaking, Catherine Haney, a senior member of the agency, handed me a piece of paper with the results of the latest simulations, the ones showing the radiation would exceed the levels deemed safe in the United States out to *at least* fifty miles from the reactor. I communicated this to the participants on the call. As we continued to discuss this situation, additional analysis showed impacts at distances of thirty or forty miles. I asked whether one of those was a better number to use, and the answer was that it didn't matter. All three indicated extreme hazard.

At this point I still clung to the belief that the best and

most reliable information would come from the experts closest to the problem: the nuclear safety regulators in Japan. "I really wish we were looking at a piece of paper from the Japanese showing their dose projections," I told the others. But most of the equipment used to assess the degrading conditions of the reactors was down. The Japanese too were flying blind.

"Would we act based on the information they gave me if this incident were occurring in the U.S.?" I asked my staff while still on the call.

The answer from those around me was a calm and succinct yes.

So we decided to go with our best educated guess. The agency's public protection mandate dictated that we keep Americans in Japan from harm, and so I recommended that the U.S. government advise citizens to remain more than fifty miles away from the accident site. After reaching this agreement, we sent the data from our simulations to Ambassador John Roos to present to his counterparts.

At this point I was ready for some sleep. My diligent staff had replaced the usual couch in my office with one that opened into a bed. I called my wife, Leigh Ann, whose steady support I'd felt through the day even though I'd had no time to speak with her. I told her what had happened that night.

As we spoke, I recognized the magnitude of my public reversal. That day I had urged all Americans to follow the advice of the Japanese government. Now I was recommending that our government tell them the opposite: to listen to us and to be more cautious than the Japanese government was urging them to be. "I don't know what the morning will bring," I told Leigh Ann. "This is unprecedented. I told the Japanese government it was not doing enough to protect our citizens. This could have a dramatic impact on our government's relationship with a vital ally. I'm not sure I should have done that, but I felt like I had no choice."

"As long as you did what you thought was right, that's the best you can do," my wife said. "Now get some rest. I love you and I'll talk to you tomorrow."

With exhaustion setting in, I was resigned to the criticism that might greet me in the morning.

The next morning I prepared myself for the consequences of the previous day's decisions. I searched out someone who could fill me in on what had happened while I'd slept. It turned out that Ambassador Roos had received assurances from the Japanese that the basis for our analysis was incorrect. My recommendation had been withdrawn.

The crisis slowed for a day, and I even managed a few hours of sleep that night at my house. But when my cell phone rang early Wednesday morning, I was patched into a conference call with several senior officials from the NRC and the Naval Reactor Program, which is responsible for the design, development, and safety of nuclear reactors on military ships.

None of these individuals was easily flustered or prone to overreaction. But they were unanimous: the situation in Japan had worsened yet again. The performance of the reactors continued to degrade, and the U.S. government had received reports that people in the Japanese utility company TEPCO and in some parts of the Japanese gov-

ernment believed the situation was beyond control. Preliminary reports indicated that TEPCO was advising staff to
abandon the reactor site.

We began to formulate a concise message for emergency
response officials in the White House, having agreed on the
needed actions. We had to get Japanese officials to bolster
the personnel on-site—even if that meant placing people in
potentially life-threatening situations. And we had to reinstate the recommendation that Americans stay more than
fifty miles from the reactor site. Finally, we recommended
that the State Department support the emergency departure of Americans in Japan concerned for their safety.

Of all the recommendations, I was least comfortable
advising that Americans leave the country. I was not a representative of the State Department and had no experience
dealing with actions affecting intergovernmental relations.
The State Department would have to determine for itself
how best to support any Americans wanting to evacuate.
But our overall message was clear: the Japanese government
needed to take additional action to prevent a catastrophe.

I was scheduled to testify on Fukushima before Congress
that afternoon. As difficult as responding to a nuclear disaster had so far been, I expected dealing with a solidly pro-
nuclear Congress could be even more troublesome.

As I waited in the anteroom of the congressional hearing room, NRC staff worked with the president's press officers, the State Department, and other agencies to prepare
to announce the fifty-mile recommendation. Around noon,
the State Department issued the press release.

Before my turn came to testify, President Obama called a meeting to discuss the status of efforts to assist the Japanese. The House Energy and Commerce Committee let me postpone my appearance so I could head back to the White House for my first meeting with the president of the United States.

In the Oval Office the president spoke authoritatively and decisively about the situation in Japan. He said he was going to call the Japanese prime minister before we took further action.

"What about the fifty-mile recommendation?" someone asked. We had, after all, just issued the press release.

The president's response was direct: "That's fine. That's a no-brainer." He was concerned, however, about the recommendation that American citizens voluntarily depart Japan. From a diplomatic perspective, this had more impact than did the extent of the radiation. The president wanted to speak with Prime Minister Naoto Kan before acting.

I spoke up. "Mr. President, I think we need to act now. We've been asking our Japanese counterparts for several days for better insight into their decision making and analyses. They don't have a better assessment than we do. I don't think we should wait any longer."

"I am aware of what's been going on," the president said curtly. "I remember being woken up at one in the morning for a briefing. But I need to speak with Prime Minister Kan before we move forward."

This was not the response I wanted, but I was glad the president had not surprisingly been following the issue closely.

———

I returned to Congress immediately after the meeting with the president. The most important part of my testimony unexpectedly came when I described the condition of the spent fuel pool in the fourth reactor at the Fukushima site.

Because reactor fuel is usually partially replenished every eighteen to twenty-four months, almost all nuclear plants in commercial operation have a deep pool of water in which to store spent fuel until it cools down, which usually takes about five years. Without the shield water provides, the number of spent fuel rods typically removed from a reactor would emit a dose of radiation large enough to kill a person standing nearby within minutes.

The nuclear safety community and the nuclear power industry have conflicting views about the hazards posed by these spent fuel pools. The safety community's greatest concern is that a pool could suddenly lose most or all of its water. This would happen only under exceptional circumstances, because these pools are very large and often surrounded by thick walls. But they are also typically built at a high elevation within plants. If a terrorist attack or a natural disaster created a hole in a pool, the water could drain. The plants in Japan had this design.

A dry spent fuel pool could then start a fire, releasing a tremendous amount of radiation, especially if some of the fuel rods had only recently been removed from the reactor. Some fuel pools have been in operation for as long as their reactors have been generating electricity; a large enough fire could release decades of radioactive materials.

Think about the last time you made a fire in a fireplace. Once the logs are burning well, a strong flow of air pulls the smoke up the chimney. If the burning object is a spent fuel rod, then instead of smoke rising into the air, it would be radioactive materials rising high into the atmosphere, from where they could be transported long distances—as far away as, say, Tokyo. That's the big danger.

This had been a theoretical concern for many years, but the damage from the earthquake at Fukushima brought it into immediate view. From the beginning of the accident, the world's focus was primarily on the reactors in units 1, 2, and 3—those in operation at the time of the accident. The reactors in units 4, 5, and 6 had been undergoing routine maintenance, during which time fuel is removed from the reactor vessel and placed in the spent fuel pool, allowing workers to inspect the equipment in and around the reactor without risk. The plant operators, however, had only recently removed the fuel from unit 4's reactor, creating the ideal conditions for a spent fuel fire.

We had already learned that the hydrogen explosions that decimated the unit 4 reactor had blown the walls and roof off most of the upper part of the building, where the top of the spent fuel pool was situated. The wreckage was devastating. Thick concrete walls were fractured or even missing entirely. Metal beams were bent. The radiation readings around the site had also spiked significantly after the explosion. We had hypothesized that this spike was caused by a number of factors: radioactive material blown around by the explosion, destroyed walls allowing further radioactive material to escape, and more. We also, erroneously, began to believe that the walls of the spent fuel pool itself had been so damaged that a significant amount of water was draining out, possibly leaving none in the pool.

As I began my testimony in front of the House Energy

and Commerce Committee, I read a prepared description of the state of the reactors. When it came to unit 4, I said the following: "What we believe at this time is that there has been a hydrogen explosion in this unit due to an uncovering of the fuel in the fuel pool. We believe that secondary containment has been destroyed and there is no water in the spent fuel pool. And we believe that radiation levels are extremely high, which could possibly impact the ability to take corrective measures."

I did not indicate that we believed there had been or would be a fire in the unit 4 spent fuel pool, but my statements still conflicted with the Japanese government's assessment. And because of an impression that the Japanese government was downplaying the situation, the media in the United States and around the world was more inclined to believe me.

This was unfortunate. I had not intended to make the condition of the unit 4 spent fuel pool a major point of disagreement; I was just stating the facts as we understood them. And the state of one spent fuel pool was in no way singly responsible for the additional actions we were advocating, such as asking that people within a fifty-mile radius of Fukushima be evacuated. But the fact that my testimony and the press release announcing the fifty-mile decision came hours apart made many people believe they were cause and effect.

Leaving the hearing room, I stepped into a gaggle of reporters who asked about the discrepancy between our assessment of the spent fuel pool and the Japanese government's.

"I hope I'm wrong," I said.

———

Many nuclear power proponents would see the Fukushima crisis as a chance to remove me from my position. There was a precedent for this. The last time there had been a nuclear crisis involving the NRC—the Three Mile Island accident in 1979—the president had replaced the chairman.

Critics of my efforts to improve safety at nuclear power plants began to investigate what they declared was a leadership crisis at the agency. Our decision to recommend evacuation within a fifty-mile radius gave them a wedge. While I was busy dealing with the international crisis, Senator James Inhofe of Oklahoma asked each of my commissioner colleagues if they had been informed of my declaration to use emergency powers. (I had not, it should be noted, been using my emergency powers.) Each responded to Inhofe's emailed inquiry by saying he or she had not been told.

This was a manufactured crisis. Nothing in U.S. law or in the agency's exhaustively detailed procedures requires the chairman to declare his or her use of emergency powers. So if I had been using my emergency powers, I wouldn't have been obligated to tell the commissioners anyway. But they *had* been informed of my actions and the agency's throughout the days following the accident. Their staffs were briefed three times a day, and I personally briefed them on a daily basis. They had as much information as could possibly be provided given the rapid pace of our decision making. More important, I had never taken any actions that went beyond my normal authority to lead the agency staff. We had simply followed standard policies and processes and determined what actions to take. There was nothing out of the ordinary about this.

I learned about this so-called leadership crisis from

Angela Coggins, my trusted advisor, and Steve Burns, the agency's general counsel, who never feared disagreeing with me. He came into my office one day around the time of my testimony to Congress to tell me some commissioners had asked about my use of emergency powers. They were concerned that I had not specifically declared that I was using them.

"Am I supposed to?" I asked.

"No," he said. "There's no formal process to declare the use of emergency powers."

"Should I send a memo or something?" I asked in the event that I might eventually need to use emergency powers.

"I don't think that's necessary," he told me.

"Okay," I said, thinking, *We can always deal with this later.* "Is there anything else?"

"There is some concern that, regardless, the emergency powers provisions of the Atomic Energy Act do not apply in this case."

"Is that true?"

"No," he said. "My predecessor clarified this issue in a memo after the terrorist attacks of September 11." He handed me the memo.

And so I dismissed the issue, not recognizing that this imbroglio would consume a good part of the remainder of my time as chairman, leading to investigations and even efforts to prosecute me for criminal violations.

CHAPTER 6

Visiting Tokyo: Crises Reveal Human Beings at Their Best

The phone rang sometime between appetizers and the main course during my mother's birthday dinner. It was rare that her birthday would fall on a Saturday, when all my immediate family could gather around her. My mom was suffering from cancer, and each birthday was another opportunity to celebrate her success fending off the disease. (She would die a year and a half after I left the NRC.) That day the phone call would end our festivities prematurely.

Flights were found to take me to London and then on to Tokyo, in a blissful sixteen hours of isolation from emails, phone calls, reporters, and grim updates about the radiation released. After just over two weeks of continuous effort to deal with the strain of the Fukushima accident, I was looking forward to spending uninterrupted hours on an airplane. Here, at least, I knew no one would ask difficult questions.

It had been over two weeks since the earthquake and tsunami, and the Fukushima nuclear reactors were still not contained. The nuclear fuel had overheated and was continuing to release radiation into the environment. Anyone spending extended periods of time near the three damaged plants would receive lethal doses. Emissions flowed into the air and water, spreading far from the plant thanks to the world's winds and currents.

The radiation continued to threaten the people of Japan and the many nations nearby. We still didn't know whether the radiation levels would increase and what this might mean for Tokyo. My job on this trip would be to better understand these issues and to improve communications between the U.S. and Japanese governments.

As soon as I landed, there was a message for me from President Obama's science advisor, John Holdren, a wonderful, intelligent man with a gray-flecked beard who, whenever he walked into a room, seemed to lean slightly to one side, as if he were on a different plane from the rest of us. Holdren had been speaking with Shunsuke Kondo, chairman of the Japanese Atomic Energy Commission and one of his counterparts. Kondo was very concerned that the situation in Fukushima could get significantly worse. He laid out a scenario that involved the complete melting of the fuel in three of the reactors and fires in each of the buildings where old nuclear fuel was kept.

Kondo believed that enough radiation could be released to cause widespread contamination in Tokyo. Emergency response actions would create a tremendous challenge for the millions of people living and working there. Holdren was adamant that I meet with Kondo to discuss his calculations.

The NRC staff was not yet predicting such dire scenarios. But the situation in Japan was very different from what nuclear analysts in America were used to. There were six reactors at the Fukushima site, a much greater number than at any site in the United States. Plus, the hydrogen explosions that had rocked the site a few days after the accident started had challenged many assumptions about the structural integrity of the plant's protective buildings. No one had ever considered the impact of an earthquake followed by a tsunami followed by a huge explosion. Chairman Kondo had started to look into the possible repercussions, and his findings were sobering. I quickly agreed to meet with him the next day.

At the airport I hopped into a car for the hour-long drive to the U.S. Embassy. During the ride, I received an excellent briefing on the reactor status from a team of NRC experts who had been in Japan since days after the accident began. Their message was clear: the reactors needed to be flooded with water so that the fuel could cool. The Japanese, however, were reluctant to do this for fear of exacerbating existing problems.

Since the reactors were leaking, water already being added to cool them down was becoming contaminated and leaking out into the ocean. The rest of the water was—by design—turning into steam and evaporating. Each drop of water that evaporated or leaked out of the buildings brought the reactors one step closer to finally shutting down. But each also put more contaminated material into the air and the ocean near the plant.

It is hard to appreciate the importance of the sea for many Japanese people unless you visit Japan and experience how confining life on an island nation can be. The sea is one of their most vital resources. And so the dilemma the authorities were facing was extremely vexing: add more water to cool the reactors and stop the airborne release of radiation but risk releasing more contaminated water into the ocean, or maintain the existing water levels to protect the ocean but prolong the disaster.

To understand the significance of this debate, we need to understand the science of radiation contamination. And there is no better place to begin than in Japan because of the nuclear bombs that were dropped on the country more than seventy years ago, exposing millions of people to intense, immediate radiation. The survivors of this devastation continue to provide tragic but valuable information about the effects of radiation on human beings. (The appendix gives more details about the physics behind this health hazard.)

Out at sea human exposure will be much reduced, but radiation can still enter the food chain as sea creatures and plants absorb radioactive particles. Carefully monitoring food coming from the ocean can successfully limit a population's exposure to these radioactive materials, but this can have a significant impact on the livelihood of those who depend on fishing to earn a living.

Before the accident, I'd believed that dealing with land contamination was far more important than managing ocean contamination. After the accident, I would learn that things were not so simple. I would come to appreciate how sensitive the Japanese were to the idea of contaminating the ocean, which was the reason the government was having trouble choosing a course of action.

As we drove to the U.S. Embassy, I realized my task was to impress as strongly as possible on my Japanese counterparts the importance of adding water to the reactor vessel. Indecision would cause more harm; we had to agree to act. Ultimately the uncontrolled release of radiation needed to be stopped.

After the challenging early response to the Fukushima crisis, Goshi Hosono had been appointed by the prime minister to lead the country's efforts. He was not a nuclear expert, but he proved quickly that he was more than capable of corralling the Japanese nuclear industry and government bureaucracy. Environmental Minister Hosono was approachable, direct, and open to novel solutions. Without his leadership the radiation crisis would not have been resolved in the time that it was.

Chuck Casto, the lead NRC expert in Japan, was my most trusted advisor there. His efforts to gain the trust of everyone with whom he worked helped make this crisis much less difficult to resolve. So when he urged me to take a hastily arranged meeting with Hosono, I agreed, asking only for a set of clear and concise talking points.

Chuck's talking points were simple: "Reactor one needs to be flooded. The Japanese government needs to stop worrying about the sea contamination and just do it." Chuck's words were actually a bit stronger than that, which is why I valued him so much. He told me what I needed to hear, not what I wanted to hear.

When Ambassador Roos and I sat down with Minister Hosono, he listened intently to what we said, then explained his position once again: the addition of water could lead to further spills into the ocean, and there was not yet any plan to deal with that. I told him we understood his concerns, but that action needed to be taken.

The words he uttered next caught me by surprise. They were candid and honest and vulnerable: "I am only thirty-nine years old."

A rush of emotion went through me. He was just one year younger than I. In my professional life, and I suspect in his, there was no one else who knew the pressure of making such profound decisions at this age. The challenge of confronting a skeptical industry and bureaucracy—made only more skeptical by our youth and apparent inexperience— was something he and I could truly understand. In one sentence he had established a connection between the two of us that thousands of miles of ocean, two separate languages, and centuries of cultural differences could not unbind.

The uncountable and unpredictable events that led Hosono and me to that meeting challenge my analytic view of the universe. I had never planned to become chairman of the U.S. Nuclear Regulatory Commission; it had simply happened. I suspect that Hosono followed a similar course of accidents and reactions. But this confluence of circumstances had created a perfect and necessary encounter.

So, feeling completely exposed, I told Hosono, "I too am only forty years old." I smiled as warmly as I could. "I know what it feels like to be in this situation. I know it's not easy. You have to do the best you can. You have to decide the best way to move forward, and know that you can do it."

He promised he would make the best decision he could.

With that, Roos, always the expert diplomat, lightened the mood by joking, "Well, I guess Chuck and I are the old guys in the room." And so the meeting ended, leaving me with the strong sense that Hosono and his team were more than capable of bringing the crisis to an end.

Up to this point I had not worried that Tokyo would be affected by fallout from the Fukushima reactors. This was one time when I may have trusted the judgment of my NRC colleagues too easily. Most NRC staff believed the distance between the reactor site and the capital was simply too great to raise concerns. We would soon learn, though, that the predictions of the nuclear industry and regulators and other government officials about the release of radiation were either quickly done and imprecise or slowly done and more precise. Neither option allowed for making the best decisions.

It may seem surprising that countries like Japan and the United States, which can design mobile phones thinner than a deck of cards, would not yet have the technology to reliably assess the effects of a nuclear reactor accident. But modeling the release of radiation in real time is extremely challenging. It depends on the condition of the reactors, the amount of radiation being produced, the circuitous path the radiation must take to escape the leaking buildings, and the atmospheric conditions around the plant and maybe even hundreds of miles away. All these factors would need to be input into a number of computer programs, using assumptions to fill in information gaps about reactors' conditions and weather patterns. Despite the computers' considerable processing power, they would still spew out their answers only hours or days later.

To put this challenge in perspective, imagine blowing

pencil shavings into the air every day for a year. It is unlikely that those shavings will wind up in the same spot every time. Think of all the things you would need to know to predict where they will end up. How hard did you blow? Was there a wind blowing in your face? Was your hand cupped or flat? How many shavings did you have? How large were they? How high above the ground did you hold your hand? How far from your hand was your mouth? Did you take a deep breath or a shallow one? Modeling the release of radiation after a reactor accident is even more complicated, but the questions are roughly the same.

President Obama's science advisor, John Holdren, had warned me of the ominous predictions of his Japanese counterpart, Shunsuke Kondo. So I was not shocked to hear Kondo tell me that if the situation at Fukushima worsened, all the reactors could leak more, the stored reactor fuel could catch fire, and his calculations showed that this could lead to contamination reaching Tokyo. But at that moment there was nothing I could do other than assure Kondo the U.S. government would know of his concerns. We had given our recommendations to the Japanese government, and the actions they were taking were likely all that could be done. Now we would have to watch and wait.

In the weeks and months that followed, as I pondered the implications of the Fukushima disaster, I would think back to my meeting with Goshi Hosono, which reinforced for me that human crises are dealt with by human beings, whose flaws, self-knowledge, and opportunities all come into play.

CHAPTER 7

Tsunamis in the Heartland:
A Scenario for an
American Fukushima

Could a tsunami ever threaten a nuclear plant in the American heartland? No, but that's not the point. It's true that tsunamis do not barrel through the farmland of Nebraska or Kansas. Fifty-foot-high waves do not crash onto the fences surrounding nuclear plants in Minnesota or Wisconsin. But the tsunami in Japan was simply the natural disaster that exposed certain weaknesses in nuclear power plant design and operation. Other kinds of natural disasters could reveal other fatal flaws.

In the United States, nature has prepared many recipes for destruction: rivers can overflow their banks; earthquakes shake the ground; and tornadoes, mudslides, and hurricanes each have their own ways of setting off a bad sequence of events inside a nuclear power plant. These disasters may be getting worse. Tornadoes in particular appear to be more frequent and severe than in the past. And hurricanes, too, are predicted to increase in intensity as global warming continues.

Most plants in the United States are designed to respond to a conservative range of natural furies. These so-called external hazards may not by themselves damage a plant, but they can disable the safety systems that stop accidents from starting. This, of course, is what happened at Fukushima. Although the Japanese government still won't know for decades the exact damage the natural disasters caused to the reactors' internal parts, we do know the earthquake wiped out the external power supply, and the tsunami disabled the internal ones. The decay heat of the reactor, unleashed and uncontrolled, did the actual damage.

To stop a deadly domino chain before it begins, plants in the U.S. must predict the effects of the worst natural hazards that could realistically happen. These assessments are difficult to complete because many of the people involved in the plants' early design lacked precise information on natural disasters that were historically likely in their region of the United States. Storms may also increase in intensity over time, while shifts in the earth can create new seismic hazards that did not previously exist. All this complicated geology, meteorology, and physics must then be fed into a computer program to predict the impacts of these events on the plant.

So while a tsunami is not likely to strike a plant in Illinois, a tornado could disable the same safety systems that the earthquake and tsunami destroyed at Fukushima. Being tsunami-proof does not make a plant accident-proof. The forces of water, earth, wind, and fire can be unleashed anywhere. This is what the Fukushima accident should have taught the United States.

Since the beginning of the nuclear era, nuclear plant designers have promised to corral, curtail, and command nature.

In 1971 the predecessor to the Nuclear Regulatory Commission established a set of standards, known as the General Design Criteria (GDCs), to maintain the safety of each nuclear power plant. The second of these criteria requires that plants be designed to handle the most significant natural danger that has historically occurred in the area around the plant:

> The design bases for these structures, systems, and components shall reflect: (1) Appropriate consideration of the most severe of the natural phenomena that have been historically reported for the site and surrounding area, with sufficient margin for the limited accuracy, quantity, and period of time in which the historical data have been accumulated, (2) appropriate combinations of the effects of normal and accident conditions with the effects of the natural phenomena and (3) the importance of the safety functions to be performed.

There are few requirements in the NRC's canon of rules and regulations that provide this sort of unambiguous and conservative direction to the designers of nuclear power plants. In contrast, the regulations dictating where a company can build a plant are much less restrictive, allowing plants to be built near but not too close to population centers. (Of course once a plant is built, the NRC can do nothing to prevent population growth around it.) In the weeks after the Fukushima accident, I would turn to this section often to ensure I knew how to answer questions about whether American nuclear power plants could withstand a large earthquake.

The answer to this question is, of course, complicated. Translating the simple English of the GDCs into the complicated language of computer codes, engineering standards, and plant architecture is an art, not a science—one that often muddies the all-or-nothing clarity of the GDCs. Power plant owners, government officials, and the scientists and engineers that each employ may not always be able to identify or imagine the "most severe of the natural phenomena" at a particular site. And what's crystal clear on paper is smudged through practical application in an operating nuclear power plant, especially one that has been online for years.

I was confronted with the limitations of the GDCs on August 23, 2011, when the earth started rocking in Washington, DC. As I was sitting in a conference room on the eighteenth floor of the NRC's headquarters, learning about the safety of nuclear power plants in the Southeast, I felt the building sway like a cruise ship in a modest swell. It turned out that an earthquake—5.8 in magnitude, which is tame by West Coast standards—had hit an area of Virginia just miles from the North Anna Nuclear Generating Station.

Attuned after Fukushima to the concerns earthquakes pose to nuclear power plants, I quickly asked for an update. The operators observed no immediate damage, but they did identify slight movement in several of the large concrete cylinders housing spent nuclear fuel. Within the hour the White House situation room informed me that staffers would brief the president, who was on vacation, on the earthquake and its impact on all types of infrastructure, including nuclear plants. Over the phone I told the president, who was on a golf course somewhere, that there had been no reported damage to the Virginia plant and that it had shut down safely.

Still, the earthquake had exceeded (although only slightly) the most severe magnitude envisioned by the designers, exposing a problem that the post-Fukushima NRC review would reiterate: the science used to model earthquakes and the resulting behavior of nuclear power plants was woefully outdated. Developed in a time when computers less powerful than a smartphone took up entire rooms, these codes were decades old, painting at best an incomplete picture of plant performance and at worst lulling us into a false sense of safety.

The shaking caused by an earthquake is a fingerprint of sorts. Much as light can be broken into the various colors of the rainbow, earthquakes shake objects with a combination of different frequencies; some earthquakes shake objects more rapidly than others. This matters because different components of plants' safety systems respond to different types of shaking. Electronic equipment and motors are more susceptible to high-frequency vibrations. The structures that physically support the plant are more susceptible to large, pendulum-like motions.

For North Anna, the deviations from the "design standard" earthquake turned out to be due to the actual quake's high-frequency shaking, which left the plant's base structure unaffected. Fortunately, this difference did not significantly impact the plant's safety, although a more severe earthquake could have. And safety should depend on more than luck.

———

A far more significant demonstration of the limits of the GDCs took place the previous spring, when melting snow sent massive amounts of water flowing toward Nebraska.

Like a giant water slide, the northwestern quarter of the United States drains melting snow down a series of natural water troughs and artificially modified rivers and dams until it all converges into the Missouri River. In the spring of 2011 record amounts of snow in the Rocky Mountains and large spring rains made for historical water levels. The U.S. Army Corps of Engineers, a civilian and military component of our armed services, controls a series of dams along the waters leading to and including the Missouri to ensure the river can be used effectively by farmers, outdoor enthusiasts, and barge shippers. That spring there was another factor to contend with: the Fort Calhoun Nuclear Generating Station.

One of two Nebraska plants sitting alongside the Missouri River, Fort Calhoun is a pressurized water reactor that began operations in 1973. Thirty-five miles south of the plant is Omaha, the largest city in Nebraska, with a population of just over four hundred thousand. That summer the city would watch anxiously as the river levels crept up to and over previously recorded flood levels—especially just outside the Fort Calhoun plant.

As the Fukushima accident showed, flooding can disable many systems at a nuclear power plant. Coincidentally the NRC had actually flagged a concern about flooding hazards at the Fort Calhoun site the previous fall, six months before Fukushima. Even as early as 2003 evidence showed that the risk of flooding at Fort Calhoun was worse than what had been planned for. But the plant owners never redesigned the plant or changed their procedures to address the higher risk. The plant continued to rely on a primitive method of flood control: stacking sandbags on narrow ledges.

Had the plant not finally upgraded its flood defenses after the NRC's 2010 inspection, operators would have had to stack sandbags more than five feet high to withstand the rising waters the next summer would bring. But five feet of sandbags perched on a narrow threshold would not have held back the Missouri. Imagine leaning against a pile of sandbags heaped on a surface less than half the width of a balance beam. You would not stay upright very long. With a large enough flood, the plant could have had an accident similar to Fukushima's; off-site power would have been lost and the on-site backup power systems would have failed. Fort Calhoun's owners should have recognized this.

This episode identifies one of the inherent challenges of the regulatory system, one that is not easily corrected: it will never be possible to inspect every operator, analysis, procedure, or component of a plant. And so the agency's oversight involves a degree of informed judgment. Resident inspectors, stationed at each plant, follow the plant's functions and performance on a day-to-day basis. The NRC takes their on-the-ground insights and integrates them into the technical analysis performed by engineers and scientists at the NRC's main offices. The agency then develops detailed inspection plans that focus on areas most in need of review. But these are just a small subset of all the possible sources of failure.

In the end the primary responsibility for the safety of a nuclear power plant rests with the plant's owners and operators. They have far more opportunity to spot weaknesses than NRC inspectors ever will. So it is troubling when the agency identifies an issue that has been evident but unaddressed for a long time. While in the end the system worked at Fort Calhoun because the potential for flooding was identified, it also failed because the NRC, not the owners, had to find the problem. And the historic Missouri flooding was about to test the temporary safety improvements the NRC had required in response to the 2010 inspection results.

So in late June 2011, I found myself standing in front of TV cameras, explaining to a nervous public why this plant was not going to have a Fukushima-style meltdown. Whatever credibility I had gained through my response to the Fukushima accident was about to be tested.

Scenes of flooding and devastation around the Fort Calhoun plant were filling American television screens. It became apparent that I had no choice but to visit and offer my views on the status of both the Fort Calhoun plant and the Cooper plant eighty miles downriver. Besides, I wanted to see for myself whether the assurances of stability I was receiving were accurate.

Just before departing I had the first indication my visit could become a problem. To make the job of the Fort Calhoun workers easier, a giant berm had been placed around the plant. This massive water-filled inner tube made entry into the facility easier by providing a dry buffer that extended out from the plant walls, allowing the use of the plant's normal doors. It also provided a powerful image of protection. But early on the morning of my departure, I awoke to a message from a senior manager at the agency. A forklift had inadvertently punctured the berm, allowing the Missouri River to rush up to the walls of the plant. As an added complication, the forklift had apparently been doing work to prepare for my visit. My staff was concerned that I would be held responsible for the incident and wanted to confirm that I still planned to make the trip.

I did, but I visited the Cooper Nuclear Station first. Cooper was the control group of sorts in my tour of the Nebraska plants. As part of the agency's efforts to communicate to the American people what was going on, I brought a contingent of print and broadcast reporters along. The visit to Cooper must have seemed to them a boring preamble to the drama they expected at Fort Calhoun. Just like Fukushima Daiini, located just a few miles from the damaged Fukushima Daiichi plant, Cooper was poised to weather the flooding much more easily than its neighboring plant, having been built at a much higher elevation. In fact Cooper was completely dry, the floodwaters nowhere to be found.

The contrast to the poorly chosen Fort Calhoun site was stark, raising once more the obvious question so many people asked after Fukushima: Why did they build the plant there? Part of the answer is that the nuclear industry simply insists that severe accidents are not possible. At Cooper everyone came across as sincere, direct, and committed to a well-planned response to whatever flooding might come. But I couldn't help but believe that underneath all that at least a few of them were thinking, *This is overkill. Our site is never going to flood.* They may have been right on that day, but there is always the potential for *something* to trigger a catastrophe.

In the Fort Calhoun plant, water was everywhere. I started by meeting with the Army Corps of Engineers' Colonel Robert Ruck, who was responsible for the dams along the Missouri River, the heart and soul of daily life for many in these parts. Listening to him reminded me of the many assurances I'd heard from people in the nuclear industry over the years.

"The dams won't fail."

"But what if they do?"

"They won't. We inspect them. They're solidly built. They won't fail."

As Colonel Ruck went through his litany of guarantees, I felt an awful sort of familiarity. Of course safety officials prefer the certainty of actual events to the discomfort of hypotheticals. The best I could say during the press conference later in the day was that the NRC and plant employees were watching the plant closely. Protective measures were in place, and we believed the plant was safe given the current water levels.

Our next step was to see the flooded Missouri. Led by officials of the Army Corps of Engineers, we boarded a helicopter to survey the damage, rising out of Omaha and quickly heading upriver. Soon it was hard to visualize the massive pool of water as a river anymore. In many places the once crisp lines defining the riverbanks resembled dashes, where water had crashed through earthen berms on both sides. From above it looked like a shallow inland sea.

The Fort Calhoun plant looked like a castle surrounded by an unbounded moat. It was no longer on the banks of the

Missouri River; it was nearly in the middle of it. The failing berm and the rising waters had forced the plant to begin producing electricity through its diesel generators, because the electrical towers, capacitors, and transformer that connected the plant to its off-site power providers were on the verge of being flooded. This created a new vulnerability, as the generators needed fuel and cooling to run. Any disruption would disable them, potentially causing a blackout. Fortunately this didn't happen; the Fort Calhoun plant ran on diesels for a long time. But to be in such a state—while far from a catastrophe—was no cause for comfort. The potential for a major nuclear crisis was there, flowing outside the doors of the plant just inches below danger level.

We returned to the site in cars, parking a fair distance away because the closer parking lot was flooded. Just below my feet, under the aluminum catwalk leading to the plant, I saw a ferocious flow of force. This was not a stream bubbling through a pastoral forest; this was a rush of destruction.

Stepping over the threshold into the plant, I sensed the calm and confidence of the employees. The isolation of the reactor buildings brought a sense of peace. Everything felt more secure; you could neither see nor hear the danger outside.

But the Fort Calhoun plant had serious problems, including the missed opportunities to fix its flood-reload weaknesses. The flaws turned out to be so significant that after the floodwaters receded, the plant would undergo the most rigorous inspection regimen the agency could conduct. That June day I did my best to convey the personal responsibility of each person for ensuring the plant's safe operations. The NRC cannot do it alone; mistakes must be caught and corrected by a plant's workers and owners. Fort Calhoun was just beginning a long journey to improve its ability to do that.

At the press conference that followed, I reiterated that the plant was being closely monitored and that it was safe. But I could not help thinking, *What if we miss something? What if a dam fails?* As chairman, it was hard to say those things out loud—but the Fukushima accident made them inch closer to the tip of my tongue.

Because the flooding at Fort Calhoun took so long to subside, there was plenty of opportunity for worry. But when it was all over, nearly everyone believed that once again the technology had prevailed. Proponents of nuclear power would continue to tout the robustness of American plants. Opponents would continue to raise hypothetical questions about undiscovered failures.

The power of nature is most obvious when we witness it up close. Leaving Fort Calhoun, it was the murky water swirling just inches below the plant's doors that stayed in my mind, contributing to the resolve I needed to press forward with the Fukushima reforms. These were no longer simply hypothetical scenarios. I had walked over the Missouri River to enter a nuclear power plant, about as close to an accident as you can get without one happening. It was a situation I hoped never to experience again.

CHAPTER 8

Fukushima Effects:
The Fight over Essential
Industry Reforms

The tragedy of the Fukushima accident and the dedication of the agency staff in its aftermath rekindled my *Mr. Smith Goes to Washington* innocence; I wanted to rededicate myself to championing nuclear safety, and I thought others would follow. I was wrong. The actions I took after Fukushima, and the fierce opposition they provoked, would open a chasm between my fellow commissioners and me. Determined to make good on my promise to fix the problems we found, I pushed through this conflict and achieved mild success with a series of reforms. But the biggest change—making reactors impervious to catastrophic releases of radiation—was too difficult to accomplish.

From the very first weeks of the accident, I was confronted during nearly every interaction with Congress or the media with two crucial questions. The first was "Could radiation from the accident harm people living in the United States?" My answer to this came from my gut—a no that was later confirmed by detailed measurements and calculations. The second question was "Could a similar accident happen in the United States?" Ever since my first public statements at the White House podium, my expert staff had offered reasons why the chances of a similar situation evolving in the United States were low. But nervous about delivering an absolute promise, I would couch my statements in a freight train of caveats. I did, however, feel comfortable vowing that if the agency recognized something that needed to be fixed, we would fix it.

But talk is cheap, and fixes are expensive.

About three months after the Fukushima accident, I had in hand an NRC study that could finally answer questions about the impact of the crisis on American plants. I knew, no matter what, that the chance of a nuclear accident was, is, and always will be very small. But I longed for certainty. Instead of offering assurances that sometimes seemed no firmer than the promises the Japanese nuclear industry had made about the safety of their reactors, I wanted to provide specific information about what our vulnerabilities were and how we as an agency should fix them.

The public start to all this work took place in the drab 1970s peach-accented commission hearing room on March 21, 2011, only ten days after the Fukushima accident began. The meeting was solemn. Everyone in the room knew of friends, colleagues, and strangers throughout Japan who were suffering. Bill Borchardt, the agency's chief civil servant and my official emissary to the thousands of employees in the NRC, presented the facts. The commissioners asked cautious questions and made careful statements. The audience listened eagerly. This was one of the first public assessments of the accident made by a reputable, independent organization outside Japan.

By the time the meeting started that Monday, I had repeated many times my promise to review all the information the agency—or anyone else—learned about the Fukushima accident and to consider its implications for plants in the United States. The phrase I latched onto was *systematic and methodical*. I knew the agency had been criticized for responding to the Three Mile Island accident by trying to do too much, producing a decades-long backlash about excessive regulation and agency overreach. I was determined to avoid that characterization; any review we conducted would be, yes, systematic and methodical.

Most important, I wanted the commission as a whole to endorse this review. Although I had the authority to initiate a review myself, I believed the commission was more likely to adopt the findings if it established the review at the outset. The strongest endorsement, I believed, would come from the commission announcing its review on March 21, with the world watching.

There were holdouts among my colleagues that day. Some wanted to ensure—before the review had even started—that its findings would be kept from the public. This was an old habit, the industry and its allies in the commission having long sought to control the release of information. The practice had become so ingrained that any time a power plant submitted a formal proposal that was woefully inadequate, the agency would communicate its concern privately to provide the power plant an opportunity to withdraw the submittal before the agency would be forced to reject it.

After several days of wrangling, I eventually prevailed: the report would be made public once completed. I didn't yet know that my colleagues on the commission would take another stab at slowing the report's release once it was ready. The unnecessary tussle over the report's publication would be a warning sign that actually doing something with the findings would be even more difficult than I expected.

Once the mission had been established, the next step was to pick the team. As with many other one-off efforts by the agency, the review would be accomplished as a task force, with a small group of people working largely independently of the agency's normal chain of command, a necessary choice given the commission's tight ninety-day time line; they would report their work directly to the commission.

Charlie Miller, an NRC veteran, headed up the task force. He was a perfect choice because he was set to retire from the agency once the review was complete. This gave him the freedom to follow the facts wherever they might go. Charlie assembled a top team of agency staff to assist him. Before they released their report, Charlie and his group were considered some of the best people at the agency.

Wishing to ensure their independence, I held just one private meeting with them, during which I reminded them that the NRC was the most respected nuclear regulator in the world. The pronouncements about Fukushima that I'd made at the congressional hearing weeks before were some of the strongest to come from any regulator. After America's firm but measured response to the accident, the agency was looked to by many countries for information and assurances. I suspected the task force's work would be similarly influential. So I told the team that their report was not just for the agency or even the United States; it would be read by people all over the world. I also emphasized that they had only this one opportunity to make recommendations. Whatever issues they decided were important and needed to be fixed would likely form the basis for reform efforts for years to come. "Whatever doors you close in your report will be very hard to open. You can always go back and close doors later, but you'll have a very hard time opening doors to further investigation that you close."

If the task force was under tremendous pressure, so in its own way was the nuclear industry. Germany would abandon its nuclear program as a result of Fukushima, a decision that, while politically popular, was a dramatic reversal from the government's previous position. The complete abandonment of nuclear power was not likely in the United States, but the industry could face costly and protracted changes to plants.

To alleviate some of the anxiety felt by people inside and outside the industry, we established regular public forums to discuss the report as it progressed. I hoped these meetings would serve as a buffer against accusations that the task force was scheming to shutter the American nuclear power industry. Given the prestige and expertise of the team's participants, such an accusation would be hard to make stick, but I felt it was a real concern. Having seen the commissioners already try to prevent the release of the report upon completion, I was alert to attempts to discredit its conclusions.

The public meetings nicely previewed the task force's findings, leaving little to surprise in the final report. But even after so much transparency, the report, released on schedule in July, three months after the task force started in late March, sent shock waves through the industry, its supporters in Congress, and several of my colleagues on the commission. It was as if the report had called for shutting down all nuclear power plants in the United States—hardly the conclusion of Charlie Miller's team. In fact one of the most prominent parts of the report was this statement:

> The current regulatory approach, and more importantly, the resultant plant capabilities allow the Task Force to conclude that a sequence of events like the Fukushima accident is unlikely to occur in the United States and some appropriate mitigation measures have been implemented, reducing the likelihood of core damage and radiological releases. Therefore, continued operation and continued licensing activities do not pose an imminent risk to public health and safety.

Of course the report did lay out what needed to be fixed. I was relieved that after months of uncertainty, we now knew where we stood and what we had to do to make us even safer. Now I could make more than vague avowals that if the agency identified areas needing improvement, we would act on them; now I could give specifics. And the report's proposals were sensible and accessible, so much so that comparable measures were adopted by most foreign regulatory bodies over the next six months.

Miller's task force developed twelve basic recommendations. The first was the most controversial; it called for creating new safety requirements that would take into account scenarios more challenging than the ones traditionally used in regulatory safety analysis, the "design basis accidents." Up until then, the industry had been able to avoid dealing with some accident scenarios because they were deemed too remote or too vague. The industry had also been able to avoid making improvements because the NRC had to prove that the accidents the improvements were intended to guard against *could* occur. Taking more accident scenarios into account could only help.

The report criticized the use of voluntary industry initiatives. In order to sidestep a new safety requirement, the industry would often present a watered-down version of the NRC's solution that the industry would then carry out itself with no regulatory prodding. This created a dilemma for the agency, because sometimes certainty about accomplishing a weak alternative was better than uncertainty about accomplishing a strong one. But these voluntary initiatives denied the agency its usual oversight, making it difficult to determine whether the voluntary efforts were doing anything at all to enhance safety.

One particular Fukushima-related fix fell into this category. One of the key preventable failures of the Fukushima accident was the hydrogen explosion. When concerns about hydrogen accumulation were raised decades ago in the U.S., the agency never followed through in a rigorous way, accepting instead a voluntary industry program to deal with explosive hydrogen. The task force rightfully encouraged the commission to avoid solving problems with voluntary industry initiatives in the future, and specifically recommended a hydrogen explosion prevention requirement.

Together the twelve recommendations provided a sensible approach to reforming nuclear safety in light of the worst accident since Chernobyl, one that occurred at a plant with the same technology used in many American plants. The challenge now was to turn this assessment into a meaningful display of agency authority and commitment. But the resistance began to build before the report was even finalized.

The final report of the task force almost never saw public distribution.

Even before the industry had seen it officially, rumors of its contents were circulating. As I said, given the number of public meetings the commission held in the short ninety days the task force worked, nothing contained in the report should have been a surprise. But I knew that to have any chance of making an impact, the entire document needed to be made public. Nearly every action recommended by the task force needed to be approved by the commission and then carried out via regulation or other means. The public and the industry would have an opportunity to weigh in every step of the way.

I pushed for the report to be made public without commentary, as the commission had originally decided. Some of the agency staff, responding to influence from several of my commission colleagues, decided to do something else.

Just days before the report's release, I met with one of the senior leaders of the agency, Marty Virgilio. Marty is a soft-spoken man whose calm demeanor and quiet advice proved crucial during our months of active response to the Fukushima accident. So I had no reason to doubt his assurances that the task force report would be made public with a simple memo stating it was being transmitted to the commission for consideration. As the commissioners had agreed after much wrangling, there was to be no accompanying review or interpretation of the report.

However, on Monday morning I arrived at my desk to discover that the report had reached the offices of the commissioners alongside a detailed review and evaluation from Bill Borchardt, the agency's chief civil servant. My fellow commissioners had secretly gone to Bill to add the additional commentary. I confronted Marty, Bill's deputy and the agency staff person with whom I most closely interacted, and told him to retract the report with its accompanying commentary and to transmit it as we had agreed just days earlier. I also demanded an explanation. None was given, but none was needed. I could surmise what had happened: I believe Bill, under pressure from the commissioners, told Marty to change it. Within hours, the report arrived at the commission without commentary, as the commission decided and the task force had intended.

This whole episode was similar to many I experienced during my time on the commission. The staff would quietly pulse the commissioners for their views and—for pragmatic more than ideological reasons—tailor their findings to the commissioners' tastes. On this occasion, I sensed that the more that happened, the more the report would be weakened even before it was finalized. So when the detailed review accompanied the task force report, I feared that this was an effort orchestrated by the other commissioners to weaken the findings. This surprise shook my confidence in Marty Virgilio. It appeared that he and Bill had been talking to the others and making decisions behind my back. Knowing that the commissioners favored some efforts to soften, I felt this was the purpose of the additional detailed memo. I had stayed out of the report, and the rest of the commission should have too.

Soon the public would see the task force's sensible conclusions, but the battle had taken an ugly, unfortunate turn. The trust I had in Bill and Marty—trust that had been built during tense weeks responding to the crisis in Fukushima—had burned away. They had conspired with the commissioners behind my back on one of the most important issues I dealt with as chairman. This fight was now my own, and I was determined to carry out as many of the report's recommendations as possible. To do that, I would need to use every tool at my disposal.

It was now July, and the heat of the summer only added to the tension that weighted the atmosphere inside the agency. My first sense of the reaction to the report came from my Democratic colleagues on the commission. There was little doubt that the Republican commissioners would oppose much of the report because they believed that Fukushima was a Japanese problem and that U.S. plants were sufficiently protected. But I did not expect they would give me much warning of their strategies in the fight to come. Their Democratic colleagues, however, let slip telling words.

When I gently prodded Magwood and Apostolakis for their thoughts on the report, their responses were so subtle it was as if they were saying something they wanted no one to hear. Apostolakis was the first to say it outright: "They did too much." Later Magwood would utter a similar remark. In their view the task force had recommended too many changes. After all, we at the agency all spoke proudly about the rigor of our existing regulatory system. How could this accident challenge that? The problems in Fukushima, critics of the report argued, were the result of cultural, political, and technical shortcomings unique to Japan, not weaknesses inherent in nuclear power itself. In fact some industry observers were unable to accept that the task force recommended changing *anything*.

I would see again just how much power lay with plant owners and their lobbyists. But they would execute their battle plan with subtlety and finesse. Knowing that a full-out assault on the report would be a public relations disaster, they praised the report and simply asked that the commission follow normal procedures to consider its recommendations. This was a subtle directive to agency staff to filter the recommendations through the prism of cost-benefit analysis.

The idea is simple: regulations that cost money must be accompanied by similar financial benefits. Of course, comparing real industrial expenditures with abstract safety benefits would require a Rosetta Stone, one that could equate the language of profits and losses with the subjective public harm from hypothetical accidents. The fact that such a Rosetta Stone did not exist did not matter: the health and environmental impacts of imposing a regulatory standard somehow had to be translated into financial terms. That meant establishing a dollar figure, which turned out to be $2,000—the supposed benefit of saving one person one rem of radiation exposure. With this figure available, proponents of the industry believed they could show that many regulations imposed an unreasonable cost with little or no comparable financial health or safety benefit.

But the courts had made clear the NRC does not have to abide by this cost-benefit system. They'd affirmed that the agency must impose the safety standards it deems necessary to provide reasonable assurances of adequate protection without regard to money. The courts left some wiggle room, however, establishing two tiers of safety requirements: those requirements that are imposed because of "reasonable assurance" and those that are simply "enhancements to safety." The latter category could be rejected due to excessive costs. In theory the agency has the discretion to determine whether a new requirement is a "reasonable assurance" or an "enhancement to safety." But in practice many in the agency seemed to have lost sight of the significance of this legal precedent.

The industry clearly favored interpreting the task force's recommendations as mere enhancements to safety, which would be subject to cost-benefit analysis. They also portrayed the report as a secret study that had no input from industry experts and government officials outside the task force, when in fact the task force could never have completed so comprehensive a document without such conversations. Still, this attack provided the industry the opening it needed to force the report to be reviewed by the remainder of the agency's many senior managers, with whom the industry communicated frequently. The more people who looked at the report, the more likely it would be weakened or modified; editing is always easier than drafting. The industry was counting on this review to undermine the report's conclusions.

Having talked to several staffers, I was confident most senior agency officials would support the task force's work, both in recognition of the difficulties of the job and simply because the report was very good. Besides, while the modifications the report recommended could potentially cost millions of dollars, none was significant enough to force the closure of any plants. And as I wrote earlier, the first and most quoted statement in the report was a reassurance that there was no immediate danger posed by plants in the United States. Unfortunately, such moderation was not enough to quell the industry's absolute opposition. They were already working on their own alternative approaches to needed improvements and would likely press for adoption of their preferred fixes instead—voluntary modifications to their plants that would be insufficient and unenforceable.

To counter their efforts, I began a communication blitz to publicize the report as it had been written. I put out a straightforward challenge: we should review the findings in ninety days, just as the task force had developed them in ninety days. Then the industry should adopt all the approved recommendations within five years—a seemingly long time, but near light speed considering the industry's history.

The industry likewise acted quickly to mobilize opponents of the report. While they never engaged me directly, I learned of their efforts through secondhand conversations, most memorably with Congressman Fred Upton, the chairman of the House Energy and Commerce Committee, one of the most powerful panels in the House of Representatives, just before I was scheduled to appear on *PBS News-*

Hour to cap off my media blitz. A genial man who insisted everyone call him Fred, Congressman Upton had demonstrated over the years a willingness to discuss serious issues even though he shared none of my political beliefs. Guiding him through the contents of the report, I tried to show how moderate and sensible it was. It was clear, however, that he had been given a very different interpretation of it.

Upton wanted me to know he saw the NRC chairman's role as that of a moderating figure in the agency's proceedings, at best working to soften the staff's positions.

"I asked the people to write this report," I told him. "It's my job to talk about it, defend it, and pursue it. Who else is supposed to do that if not me?"

It was then that I realized I would be the sole advocate for the changes the report proposed—although I would soon learn that I still had very powerful allies.

By this point the commission was fully entrenched in the guerrilla tactics of delay. My colleagues would ask for more analysis, more reports on the report, more external experts whose opinions we should engage.

The first major blow to the report came with a vote to reject the recommendation that the agency take into account more accident scenarios. But while this was a blow, it was not fatal (the report still recommended many other changes)—although it would be if the agency staff felt similarly dismissive. This showdown would come in a commission meeting scheduled for August 2011, when the agency staff and other nuclear industry stakeholders would weigh in on the report.

In between the commission's vote in late August and the September meeting, I had time to discuss the report with the staff. Privately they supported it, and so I encouraged them to say so publicly. I also went over the questions I would ask in the open session with them. At the August meeting, I then asked the same questions I had posed in private. Their answers were consistent, giving the whole scene the feeling of a table read of a script. I read my questions; they answered with little or no emotion. The commissioners were incensed because I had outmaneuvered them and the staff had agreed with me publicly. The commissioner and the industry had lost one of its greatest political tools: an ambivalent NRC staff. So they accused me of having bullied the staff into agreeing with me, but I had done nothing more than ask questions and prepare them.

Over the next several months the commission would have more votes on delaying, diluting, and denying the report's main points. Reform became less and less likely. Industry officials, eager to win a moral victory, or at least appear to do so, proposed alternative safety measures, such as deploying a set of power supplies, water pumps, and other equipment to better respond to severe accidents. But these were half measures, not meaningful changes. For one thing, they would be carried out only voluntarily, which had been the main problem with safety upgrades proposed before now.

In the spring of 2012, with the first anniversary of Fukushima fast approaching, the commission did at last recommend the quick adoption of three safety measures. Plants with a design similar to those at Fukushima would need to place under regulatory review their existing systems to manage the release of pressure during accidents. They were also required to develop interim solutions in the case of a total loss of electric power. And they would need to develop better equipment to diagnose the condition of their reactors' spent fuel pools.

These three measures were the first steps in what should have been a forceful response to a nuclear power plant disaster involving American nuclear reactor technology. Instead they were bandages to cover the problems identified by Fukushima. Real changes involving lasting protections against floods, earthquakes, and long-term power outages would take years to develop and finalize. But with time the momentum for reform would be lost. The industry's stamina would outlast the energy of reformers inside the agency. And with my five-year term as commissioner ending within the next eighteen months (and therefore likewise my time as chairman), the industry figured they could wait until I left to gut the remaining safety measures.

CHAPTER 9

Express Lane: The Nuclear Industry Licensing Juggernaut

A pause. A break. A respite. This is what should have happened after Fukushima but did not. After the most significant accident ever to occur at a nuclear power plant based on an American design, the industry's and the agency's efforts to complete the licensing of a new nuclear reactor never stopped for even one beat. Instead of a renewed focus on safety and an effort to truly understand the implications of the accident for the safety of American reactors, the drumbeat just continued: Don't stop. Don't stop. Don't stop.

The pressure to license was intense because the nuclear power industry had failed to maintain a steady construction program for new plants. In the late 1970s and early 1980s, the problems exposed by the Three Mile Island accident, the decreasing demand for electricity, and massive construction failures led to the cancellation of over one hundred unfinished nuclear power plants. No company was willing to assume the financial risks of building a new plant and so the industry plateaued at 104 plants in the late 1990s. But in preparation for the plants' forty-year licenses to expire in the next several decades, bringing an avalanche of closures, the nuclear industry began a fresh push to license and build new facilities to keep the industry alive well into the future. The timing seemed ideal—until the Fukushima accident happened.

At the dawn of the commercial nuclear era, the United States was the primary supplier of nuclear technology to countries outside of the Iron Curtain. Today most reactors in use around the world either use American technology or are derived from it. Yet over the decades, as dozens of countries pursued nuclear energy, the promise of nuclear power lost its luster in the United States.

Some industry supporters partially blamed this stalled development on the Nuclear Regulatory Commission. Until the 1990s the licensing of new reactors occurred in basically two steps. First, companies proposed enough of a design to get construction approved. Then, as the plant was being built, the owner would finalize the design. Upon completion, the owner would obtain a license to operate the facility. In the worst-case scenario, a project could fail to obtain an operating license after it had already cost billions of dollars to build. No energy company was willing to take such a risk, especially given owners' repeated failure to build plants on time and within budget. And so while the United States continued to boast the largest number of commercial nuclear power plants in the world, as other countries built new, modern reactors, American reactors started getting old.

Reacting to industry pressure, the agency attempted to address these concerns by flipping the construction and licensing process on its head. Instead of building first and licensing second, the agency would issue a license to operate a plant based solely on the planned design. The NRC established a series of inspections, tests, and criteria to ensure the plant performed in practice as the reviewers expected on paper.

The revised process dramatically curtailed the public's ability to challenge operation of a plant. Now the only way to launch such a challenge was at the beginning, during the license review. For projects that could take decades to build, this meant that people living near a new plant might not even have been residents when the license review took place. But the companies were pleased, as they could build a new reactor with a lot less financial risk than before.

A number of applications for new reactors came before the NRC just as I started working there; the industry called it a Nuclear Renaissance. Wall Street committed financing, and the agency channeled its resources toward evaluating the designs. Then Fukushima happened. The promise of reforms and the unknown costs for plant changes threatened to derail the decade-long effort to finally bring at least one new nuclear plant onto the electricity grid. The industry was determined to fight back.

In Washington today there is no more powerful statement than one that includes the word "job." When lobbyists want to influence a member of the House or Senate, they say, "This bill will bring hundreds of jobs to your district" or, an even more powerful statement, "This proposal will kill hundreds of jobs in your district." When the nuclear industry conveyed that message to Congress, they were taken seriously. This is because most nuclear power plants in the United States are in small towns, where the taxes they pay to local governments support the construction of new schools, roads, and fire stations.

Most nuclear power plants also provide hundreds if not thousands of permanent jobs, not to mention thousands of well-paying but shorter-term engineering and construction jobs. Operators, who control the reactor, can make six-figure salaries in communities with a very low cost of living. They live in homes near the plant, send their children to local schools, and shop in neighborhood stores. When their neighbors raise safety concerns, plant workers rightly argue that the issue is personal for them too because their families are just as likely as others in the community to be affected by an accident.

I never imagined that just two years into my chairmanship, I would face the question of whether to suspend work on new reactor licensing. At the time of the accident in Fukushima, the agency was in the late stages of reviewing the license applications for four new reactors in Georgia and South Carolina. Both used technology derived from the remnants of one of the great American energy companies, Westing-house.

A long and expensive process had brought these designs to within a year of the finish line. The new license-first-build-second process was meant to make nuclear reactor licensing as simple as a Lego set: take an approved design, attach it to an approved site, then complete a limited licensing review. All this was to be done before any significant construction took place. The industry wanted assistance from the NRC throughout this process, so that when the time came to win approval for a license, few problems would arise.

Unable to control their urge to tinker, the engineers who had designed the new reactors were constantly revising. The Westinghouse AP1000 design was on its seventeenth revision by the time the NRC made its final decisions. Despite a decade of efforts to streamline the licensing process, the industry still lacked the discipline to take advantage of the improved system. This new process was built around the assumption that design issues would be approved and finalized once. But each time they changed the design, the agency would have to review it.

Another stumbling block to building a nuclear power plant was the cost. At that time, new plants were expected to cost about $5 billion per reactor. That meant a two-reactor site—the plan for both Georgia and South Carolina—would require a $10 billion investment. Most of that would come from loans. Given the nuclear industry's dismal track record for controlling costs, banks had been willing to lend only at high interest rates due to the significant risk of project default. In 2010, however, the federal government agreed to back these loans for the Georgia project. Because American taxpayers were guaranteeing the loans in the event of a project's cancellation, financial firms were expected to finance these projects at a much lower interest rate.

Publicly the industry was putting before the NRC eighteen applications to build about thirty new plants. These were statistics I routinely quoted in the months before Fukushima, though I knew that only a few reactors make it past the press-release stage. And as I knew from private conversations with financial institutions and industry leaders, Wall Street was willing to finance only four to six reactors. The so-called Nuclear Renaissance would occur only if these first projects were built on time and on budget.

Even before Fukushima, the fate of the new reactors was as precarious as a bear walking on a tightrope; after the accident it was like an elephant trying to balance on a spiderweb. With so little margin for financial error, the industry would rely on its political influence to get the elephant safely across.

Just weeks after the Fukushima accident began, I convened a meeting with my colleagues to discuss the plan for new reactors in Georgia and South Carolina. I asked if we should postpone our licensing review. There certainly was a model for doing so; Japan, Germany, and France would soon change the course of their nuclear power programs. But the answer I received was no. The cautious actions of those nations stood in stark contrast to the bulldozer mentality of the American nuclear power industry and the majority in Congress who supported it.

Under the Atomic Energy Act, the public has the right to formally object to a license, but in the decades since this provision was enacted, the commission has slowly whittled down the language to make it more difficult to do so. The owners of the Georgia and South Carolina plants succeeded in overriding any formal objections to their licenses by concerned citizens and public interest groups. The scope of the licensing review is massive, but most of the technical oversight was outside the formal hearing process. So the public interest groups focused on negative environmental impacts of the plants and improper processes during the safety review. In all cases, they were unsuccessful. But this is no surprise, because the system is designed to make it almost impossible for these outside groups to succeed. There is a residual echo in the law, however, that requires a mandatory public hearing led by the commission, especially when there is no outside challenge.

In my early years on the commission, I had helped shape the scope and size of these hearings as the commission prepared to hold them. Now I was in a position to conduct the first one in the history of the NRC. I wanted the hearing to demonstrate how seriously the agency took the issue of new reactor licensing.

When it came time for the hearing on the Georgia reactors, the room was full. Hundreds of people, all of whom could potentially be called as witnesses, needed to be sworn in. The packed audience stood together and took an oath to truthfully answer any questions the commissioners might ask.

For two days we were supposed to probe, investigate, and confirm or deny the basic facts of the plant owner's statements and the agency staff's determinations. To me, the biggest outstanding issue was the unfinished response to the Fukushima accident. Should these new plants be made safer or be required to document their ability to better withstand a Fukushima-style accident? I was prepared to ask detailed questions about specific items in the thousands of pages of documents. In both the Vogtle (Georgia plant) hearing and the subsequent V. C. Summer (South Carolina plant) hearing I focused on how to ensure that these new plants were prepared for a Fukushima-style accident. The agency staff assured me they could provide a reasonable approach. I would learn later that the answer was not so simple.

It was disturbingly obvious, however, that my fellow commissioners had done little to prepare and were not interested in backing my concerns.

I soon learned they were focused on another issue.

I was tired. Within the span of two weeks in September and October, the commission had held the two-day mandatory hearing on the Georgia nuclear plant, a public update about the Fukushima reform efforts, and a two-day mandatory hearing on the South Carolina plant. Finally, after the last session concluded, I had a chance to catch my breath and catch up on the news. I shut my office door, opened a Styrofoam container with a tuna fish sandwich and hard sourdough pretzels, and began to chomp, read, and watch. Then my chief of staff, Josh Batkin, came in.

Josh was one of the first people I hired when I started at the agency. He was a tough negotiator, a kind father, and a loyal friend. He kept a baseball hat in his office with the phrase "Young Bureaucrats" embroidered on it. He is an exemplar of what public servants should be.

During difficult moments I would joke with Josh, "What's the worst my colleagues can do? Send a letter to the president telling him they're unhappy with me as chairman?" That day Josh was holding a large yellow envelope. Inside was a letter signed by all four of my colleagues on the commission detailing their concerns with my leadership style. I knew there was tension, as I could tell from the near-weekly individual sit-downs I had with commissioners or the frequent meetings with the whole commission. Still, I was surprised when the letter arrived.

I admit that I sometimes behaved in a way that could be described as hotheaded. I challenged the bureaucracy, pressed for progress, and never shied away from confrontation. I constantly worked to temper any excess, but was never as perfect as I wanted to be. And I believe that without these traits I would never have had the strength to persevere against a relentless, powerful nuclear industry that preferred having a chairman who never publicly discussed its problems.

Over the next several months the commission continued to do its usual work. We met, voted, and deliberated a number of significant issues. My colleagues had not yet made the letter public; they had sent it directly to the president's chief of staff. But I knew from my own congressional contacts that many people in Congress had a copy of it. Soon the public would too.

After a number of meetings with the president's new chief of staff, Bill Daley, it became clear to my colleagues that the president had no intention of replacing me as chairman. That left the commissioners with only one recourse: to seek the assistance of Republicans in Congress. They turned to a bombastic representative from the state of California, Darrell Issa. Congressman Issa was possessed of a fierce gaze, a thundering voice, and a wicked determination, which, in the service of his position as chairman of the House Oversight and Government Reform Committee, made him a formidable opponent.

In the years Issa ran the committee, which could investigate almost any agency in the federal government, he launched broad, prosecutorial investigations into many. As I watched his aggressive inquiries into allegedly serious violations of the law or public trust, I never imagined he would soon go after me. But, as I later learned, the commissioners, through their personal staff, had been in regular contact with Issa with complaints about me long before they sent him their letter. After a first probe found nothing worth investigating, the Democrats on the committee asked my colleagues to put their concerns in writing or stop complaining, thinking that they would stop. But they didn't. When the president refused to act on their formal complaint, Chairman Issa had no choice but to hold a hearing. With no real transgression at the heart of my colleagues' grievances, Issa's only angle would be to force me to sit through hours of testimony and questions related to my personal faults. I was not prepared, however, for the things my colleagues would say.

On the eve of the hearing in mid-December, I gathered with my personal staff and the senior leaders of the agency. We met as we always did to prepare for congressional testimony, discussing possible questions and how I would answer. This instance was no different, with one exception: how I came across would be just as important as what I said. With my colleagues describing me as an out-of-control bully, the Republicans were eager to provoke that kind of behavior in me. I suspect they believed that with very little prodding I'd transform into the Incredible Hulk. To put me in the softest possible light, my unflappable congressional affairs director, Becky Schmidt, suggested I wear a brown suit: "Brown makes you look less intimidating."

There I was, wearing my kinder, gentler brown suit, sitting in the middle of the long congressional hearing table with my commission colleagues on either side, staring at the terraces of seats on the congressional side of the dialogue. To my right were the Democrats. To the left were the Republicans. Because this was not a Hollywood version of a House hearing, members would drift in and out.

The process started with statements from my colleagues, whose words shocked me. Commissioner Bill Magwood claimed—offering no specifics—that I was a special kind of bully. "In my experience," he said, "the behavior I've heard of from some staff is consistent with a person who is abusive to women."

I was stunned. I'd kept reassuring myself in the weeks leading up to this hearing that the worst I could possibly be accused of was yelling at people.

When it came time for Republican Trey Gowdy, a former prosecutor, to take over the gavel, he attacked me as a prosecutor would, at one point threatening me with jail time. "What do they call someone who did what you did—an inmate?" he asked. But what had I done? Even on my most impatient days, I had hardly committed a crime, let alone given cause for a multihour hearing.

And yet by the end of my testimony, I felt comfortable that the situation had been largely resolved. As Issa closed the hearing, he refrained from calling for my resignation. I can only believe this was because there was simply no basis for doing so. The fire-breathing villain the committee members had expected to see before them obviously did not exist.

Both Democrats and Republicans on the committee implored us commissioners to get along. Little would change in our relationships, but the hearing forced the others to accept that I would not be leaving anytime soon. To those commissioners who would speak with me, I apologized for the slights I had committed. But I refused to apologize for pursuing the measures I thought necessary for nuclear safety.

Fukushima had led me to redouble my commitment to the safety of nuclear power in America. Witnessing the devastation caused by that disaster, I had lost any remaining patience for delaying, deflecting, or coddling the nuclear power industry. We had work to do. But the industry, with their dreams of a Nuclear Renaissance at stake, struck back. The weakness they chose to exploit was my propensity to occasionally lose my cool—and they were right that I could be hotheaded and self-righteous at times. But my worst flaw was being naive to the impact my intensity could have on others.

Just a few days after the hearing, the Democratic-controlled Senate held a second hearing to give me an opportunity to address a more favorable audience. Senator Barbara Boxer, a principled, tenacious defender of equality, opportunity, and women's rights, opened the hearing with a priceless endorsement, fiercely dismissing the allegations of mistreatment of women at the agency. Her words gave me the strength and courage to continue to fight for the causes I believe in. But despite her eloquent intervention, my reputation had suffered a major blow.

CHAPTER 10

Cleanup Is Forever:
Visiting Fukushima

In December 2011, I returned to Japan. In the months since the accident started in March, the situation had improved. The reactors were no longer hot enough to boil the water cooling them and the radiation releases into the atmosphere were no longer a serious hazard. As a result of the improving reactor condition, I was there to formally declare that the crisis was over. The announcement would happen in a series of carefully planned formal gatherings with the Japanese officials in charge of the accident response, including Minister Goshi Hosono. There was no way I could visit Tokyo, however, without first taking a side trip to see the damaged nuclear power site at Fukushima. After so much time following the accident from afar, I needed to see what the reactors looked like firsthand. The visit would reinforce for me the fact that the debates I held in office buildings thousands of miles away mattered. The destroyed reactors were a stark testament to the importance of the work I had been doing and the need to persevere.

I was humbled to finally see the devastation that the earthquake, tsunami, and reactor accident had brought to the area surrounding the Fukushima plant. Like every visitor, I had to don a white Tyvek suit, made of the same material used to wrap houses to protect them from moisture and dust. To protect my face and lungs I wore a respirator, a thick plastic mask with a clear shield that fit snugly over my entire face. The mask pulled tight against my cheeks, so tight that I could feel the rubber seal digging into my cheek-bones. To complete the outfit, my hands and feet were wrapped in gloves and booties taped to my suit. As soon as I put on the gear, every part of my body began to itch.

But no suit, no respirator, no gloves, and no booties could stop radiation with high enough energy from pene-trating deep into the body. So workers had to be fitted with special monitors to tell them when their radiation levels were approaching the limit of what they could safely with-stand. Some workers could be limited by minutes. Some by hours. At that point their work was done. They would with-draw, and the tedious process of disrobing would begin—only to be repeated day after day for months and months.

Six months after the accident, the mystery was where the spent nuclear fuel had ended up. Early models and cal-culations showed the obvious: much of the fuel had melted in reactors 1, 2, and 3. Now this molten mass was some-where within the reactor buildings, perhaps—if we were lucky—still within the reactor vessel. This is what had hap-pened at Three Mile Island. It was likely, however, that the fuel in Fukushima was severely damaged. More likely, some or all of it, especially in unit 1, had poured out through the reactor vessel into the surrounding buildings.

Despite the nine months that had elapsed since the start of the accident, water still needed to be pumped into the reactor buildings to keep the fuel cool—although on the eve of my visit, a significant milestone had been reached.

The plant owners were able to claim they had achieved cold shutdown, meaning that the water temperature inside the reactor—and therefore the fuel temperature—had dropped below the boiling point. Without the water boiling, the task of cooling the reactors down further would be greatly simplified.

But the natural landscape around Fukushima created a problem that was not going away soon. The site worked like an amphitheater, with hills cradling the reactor against the shore of the sea. This geography meant that any natural water under the ground would flow down to the reactors. Cracks in the subsurface of the reactor buildings provided a path for this water to flood the buildings, becoming contaminated, before continuing out into the ocean, creating a low-level, persistent radiation leak. Air emissions were close to nothing by the time of my visit, but even years later the problem of water contamination would continue. The accident was not over. It had simply transitioned to a new phase.

The next goal was to decommission the facility altogether. The Three Mile Island and Chernobyl accidents provided the best insights into how to go about doing that. The first and most straightforward step was to decontaminate the area around the reactor. It's a myth, however, that there's some way to magically make all the radioactive material disappear.

Radiation comes in a variety of elements, each with unique properties. Some are more easily absorbed into vegetation. Others are harmful only if inhaled; their radiation is too weak to enter the body otherwise. The weaker materials are often the longest-lasting. All radioactive materials eventually break down into natural stuff that's no longer harmful, but that process can take an extremely long time. We typically think about the half-life of a radioactive isotope, which measures how long it takes for half of a supply of radioactive material of one type to decay into something else. But the "something else" could itself be radioactive, meaning that the whole decaying process could take a very long time.

Materials with half-lives measured in seconds, minutes, and days were no longer a problem nine months after Fukushima. For those with half-lives measured in decades or millennia, the challenge remained. In the best-case scenario, cleanup means removing this material to where it poses less of a hazard to people and the environment. But in a country like Japan, land serving no useful purpose is difficult to find. Cleanup becomes cleaning one area and intentionally contaminating another, although ideally in a much more controlled manner.

But in much of the area surrounding the plant, it is simply not feasible—economically or technically—to remove all the soil, plants, and buildings that may be contaminated. The engineers can clean up only as much as costs will allow. These buildings are homes, schools, and businesses where people lived, worked, and played, and so the choice of where to spend money is gut-wrenching. Even if the cleanup is successful, people may still not want to return to their homes. Reconstructing some of the communities that existed before the accident will be impossible.

But decontaminating the countryside, despite its challenges, is far simpler than decontaminating the reactors themselves. It will be years if not decades before the engineers, scientists, and managers responsible for cleaning the site will know the exact condition of the reactors. In the years just after the accident, several attempts to send robots into areas where there may be reactor fuel failed because the robots were rendered useless by the intense energy inside the buildings. Until a clearer picture emerges of where the fuel has pooled, no one can really say what cleanup will entail. Solutions range from destroying the buildings and placing all that rubble and radioactive fuel into a specifically designed disposal area to simply leaving the fuel entombed forever in the reactors. The former is certainly the preferred choice, but the fuel may be so damaged that such a solution is not possible. So while the accident may be over, cleanup may take close to forever.

As I approached the site with its green forests amid tumbling hills, Fukushima seemed peaceful. By now the unit 1 reactor, the most damaged of all the units, was shrouded in a white protective cover that hid its mangled steel and blown-out walls. It had the look and feel of a circus tent. Our first stop was the makeshift command center for the mass of activity aimed at shoring up the unstable units and cooling the reactors. Inside the protective tent, the contamination levels of airborne radiation were nearly at a level where full-face respirators were unnecessary. This was a small measure of progress.

I was eager to traverse the site to view up close what I had seen only in videos that had become ubiquitous in the nuclear safety world. Our team got into a small bus and began a counterclockwise trip around the four reactors. We drove by a densely forested area with some of the most intense radiation still left around the site. The crews had cleaned up debris strewn about from the hydrogen explosions and gotten rid of surface contamination, but vegetation had absorbed much of the radiation, posing a challenge for cleanup.

Nearing the southernmost part of the site, I saw up close the damaged unit 4 reactor building, now a ruin of twisted metal, fractured concrete, and steel rods that rose from the gaps in the walls like the tentacles of a jellyfish. But inside was the real hazard: highly radioactive nuclear fuel that would require decades of patience and ingenious handling before it became dormant.

Rounding the slope down to the sea, we saw the ocean stretch serenely out in front of us. A small breakwater spanned part of the front of the plant, clearly insufficient to arrest the tsunami's onslaught. Remnants of massive steel tanks had been warped by the force of the sea, and a wave had shifted one large structure inland. Much of the debris and rubble had already been removed, but it was clear that the plant—built far too close to sea level—had never had a chance. No amount of bravery or engineering would have been able to stop the natural forces bearing down on the reactors on March 11, 2011.

If my resolve to make serious reforms at American plants had been strong before this visit, it was reinforced by the wreckage I saw everywhere I looked. The plant had been no match for the design flaws, poor location, and natural hazards that had converged to create lasting devastation for the people nearby and the environment. The promise of perfect nuclear safety was a mirage.

———

What about the remaining nuclear power plants in Japan? The six reactors at the Fukushima Daiichi site were permanently closed, four of them damaged beyond repair. All fifty-four of Japan's reactors would be shut down within months of the accident, reducing the country's electricity production by nearly a third. Today only a handful of plants have restarted.

This loss of electricity was catastrophic for such an industrialized society. To deal with it, the Japanese adopted conservation measures, such as setting thermostats to 83°F in the summer to reduce air-conditioning and turning to oil, natural gas, and coal. This dramatic shift in electricity production had several significant consequences, one of the most important being that the country went from successfully reducing its greenhouse gas emissions to struggling to meet aggressive targets.

The other notable consequence was the cost. While uranium-using nuclear reactors were expensive to run, the fuel itself was relatively cheap. Now Japan was forced to import significant amounts of fossil fuels. It is a testament to the determination of the Japanese people that they were able to handle this situation so well. The cost of replacing the reactors with new electricity plants—whether fossil, nuclear, or renewable—could come to $100 billion or more. Cleanup could cost hundreds of billions.

Japan also lost vital manufacturing facilities in the region northeast of Tokyo; agricultural production was devastated too. Rice and other products grown around the plant and fish pulled from the nearby ocean may contain radioactive concentrations above the levels allowed for sale. But that is almost irrelevant; many Japanese consumers would believe products from that region were unsafe even if it were proven otherwise.

The human cost is highest of all, and the most difficult to calculate. In spite of the haphazard evacuation, few people will suffer diseases that can be directly attributed to the accident. But many people have been displaced from their homes for good. Even if they could return to areas with less contamination, where limited cleanup may be possible, many businesses might not, making those communities unsustainable.

To celebrate the cold shutdown, Minister Hosono organized a gathering of American and Japanese officials who had helped achieve this milestone. But hanging over the festivities was the knowledge that the truly difficult work was yet to begin. The countryside needed to be cleaned. People needed to return to their homes. Communities needed to be reborn—whether in their original locations or in new areas. Water—endless streams of contaminated water—would have to be processed for years or longer.

During the celebration Hosono pledged that he would return with me to the site when we were both eighty, to see a new landscape recovered from the accident. Forty years on, perhaps the cleanup will be completed. And if not—if future generations are denied access due to continued hazards from radiation—the area could become a memorial to the promise of nuclear power, one best witnessed from a distance.

CHAPTER 11

Going Nuclear: Confrontation over the Building of New Atomic Plants

I returned to Washington to help decide whether or not to approve the new reactors in Georgia and South Carolina. As with so many controversies I had dealt with, issuing a final decision on the pending licenses was shaping up to be yet another showdown. I was in no mood to whack a stick at yet another hornet's nest, but when the hornet's nest is hanging over your front door, you really are left with no choice but to take it down. This decision would be my last hornet's nest.

Nine months on, the attention generated by the Fukushima accident had waned, and with it the demand for reform. Meetings, proposals, counterproposals, industry lobbying, and congressional pressure would no longer garner the attention of the media. The nuclear industry—buttressed by the campaign donations they made and their threats of job losses—would regain the upper hand in shaping post-Fukushima reforms. But I had not forgotten Fukushima.

I felt this pressure directly in late 2010, when I had a breakfast meeting with Tom Fanning, the new CEO of Southern Company, a powerful electric utility in Georgia planning to build two new reactors at the Vogtle nuclear power plant site. I rarely held meetings with industry executives outside my office, believing that these encounters fostered an air of nontransparency, but I felt dialogue was crucial, so I agreed to meet with Fanning.

I did not expect him, the head of a company undergoing a licensing review, to directly make demands of me. I expected us to engage in a dance-like ritual, with Fanning briefing me about the progress Southern was making to comply with the NRC's safety requirements and me updating him on publicly available information about the agency's pending decision. The public message I was giving around that time—which was the truth to the best of my knowledge—was that safety considerations would ultimately determine when and if the commission would approve new licenses. But over breakfast Fanning pressed for a definite answer. He threw at me economic goodies the plant would provide and, then, ignoring the usual subtle forms of encouragement and veiled threats, pressed for a definite date for when the NRC would issue Southern's new plant license. I was surprised—but even if I'd wanted to, a definitive yes was not an answer I could give on my own.

After the meeting and throughout the post-Fukushima period I continued to state that the license would be granted only if there were guarantees that the new plant would comply with existing regulations and post-Fukushima reforms. But I came up against the unwritten first commandment of nuclear policy: Thou shall not disapprove licenses. The NRC may politely ask the license seeker to withdraw an application that appears headed toward defeat, as is often done to avoid the messy situation of the nation's nuclear safety regulator actually having to veto a project. (The effect in

Washington and on Wall Street would be enormous.) But usually the applicant strives, struggles, and contorts itself to receive approval from the agency.

It may seem as though it would make no difference whether the agency rejects an application or the applicant withdraws it. But I felt we should be demonstrating to a skeptical public that safety was the agency's first priority. We should get just as much credit for disapproving an unacceptable license as praise for approving an acceptable one.

But the agency had no history of disapproving licenses, and I did not expect to reject the new reactor licenses for the Georgia and South Carolina plants. In fact, after the recent public disagreements with my colleagues on the commission, I was looking for an opportunity to show unity. But one thing kept motivating me: we had to ensure that the newly planned plants could not fall victim to a Fukushima-style accident.

A license is a powerful shield for a nuclear plant owner. Before a license is issued, the agency is free to insist that designs adhere to all parts of U.S. regulations. But once the agency has given the okay, it faces tremendous pressure getting newly found problems through "backfit" cost-benefit analysis.

The best tool for avoiding this hassle is a license condition: a commitment spelled out in the license requiring the owner of the plant to take specific actions detailed in the condition statement. Like a director giving final notes to the cast before opening night, the license condition is the agency's last opportunity to shape reactor performance.

Once the license is issued, the agency can keep the plant true to script but not edit it.

The first mandatory hearing for the Vogtle plant had taken place in late September 2011, and the second mandatory hearing for the Summer plant in South Carolina shortly thereafter. These were formal proceedings with witnesses testifying under oath and reliable information appearing as exhibits or evidence. In early October I was still optimistic that something could be worked out. This was before my colleagues sent the letter to the president criticizing my leadership.

Throughout the hearings I asked the agency staff if there were concerns about adding a license condition. I heard none; the staff could accommodate whatever the commission wanted. I was confident of my ability to convince my colleagues on the commission that a license condition was necessary. All I needed now was the specific wording that the agency staff thought would do the trick.

The new plant owners themselves were already touting the features of their plants that would make them better able to withstand the challenges of an accident like Fukushima. It seemed simple to have them commit to adopting these reforms without arguments later on about backfit. If they were not willing to certify the safety of these new designs against a Fukushima-style accident, then the commission should not be ready to move forward. That, at least, was my view.

Then came another shock: in spite of saying publicly and under oath in the second mandatory hearing that they would write up a license condition to address new safety concerns, the NRC staff now declared in writing to the commission that they no longer believed they could. I never learned why they changed their minds. I am reluctant to speculate on why they did this, because I just do not know.

My disappointment at the staff's failure to provide a license condition, coupled with the embarrassment of my own congressional hearing, began to melt my resolve. But my personal staff, who had stuck with me through that fall's accusations, were eager to help. In early 2012 they came to me with a proposal. Their idea, which they pitched with enthusiasm and conviction, was to have me propose a reporting condition instead.

This reporting condition would require the licensee to inform the NRC well in advance of the start of operations about how their plants would cope with the problems identified after Fukushima. With such information, the commission, the public, or Congress could raise concerns if little had been done in response to the recommendations of the post-Fukushima task force. I figured this was pretty close to a license condition, because if the new plants' reforms were far from complete, the commission and, more important, the public, would be given formal notification.

That was all I asked for. It was so simple, so obvious, and so reasonable that even I hoped the commission would accept it. Then I would be able to join them, the administration, and most of Congress in celebrating the approval of the first new reactor in decades.

Not one other commissioner would entertain the measure. By this point, no commissioner was willing to consider *any* proposal that I made, however modest, that would commit the new plants to additional Fukushima-inspired requirements. But I still felt I should try. I came closest to convincing Commissioner Apostolakis, the kindest and most conflicted of my colleagues. As we sat in my office discussing the license, he brought up my proposal to add the reporting requirement. "I just want to understand it," he said. "I'm not sure it does anything."

"You're right," I agreed. "It only makes them report facts. They don't have to *do* anything."

"Okay," he told me. "I'll consider it."

But even that was too much. Days later Apostolakis let me know he could not support a reporting condition, leading me to consider the only alternative I had left: opposing the license. With all the controversy surrounding my actions over the previous few months, I was not looking forward to stepping back into the spotlight by dissenting from my colleagues on the commission. And I knew one "no" vote from me wouldn't do much; licenses could be approved by a simple majority. But what else could I do?

It was my staff who ultimately convinced me, throwing back at me words I often said to them: "If we won't do this, then who will?" If I would not say that we cannot license new reactors without acknowledging that Fukushima happened, who would?

So weeks later, in February 2012, when the commission formally voted on the new reactor license for the Georgia plant, I led the meeting as chairman, cast my vote in opposition, and gave the simplest explanation I have ever offered for a dissenting vote: "There are significant safety enhancements that have already been recommended as a result of learning the lessons from Fukushima, and there's still more work ahead of us. Knowing this, I cannot support issuing this license as if Fukushima had never happened. But without this license condition, in my view, that is what we are doing."

As ever, the money, power, and promise of a new generation of nuclear reactors were too enticing for common sense to prevail. And with about a year to go on my term as chairman, I began to ponder my future outside the agency.

The time to leave would come sooner than I knew.

CHAPTER 12

Out of Time: My Departure
from the Agency and the
Future of Nuclear Power

U sually Senator Reid had no trouble getting to the point. If you were meeting in person, once he'd said what he had to say, he'd stand up and walk back to his desk. If you were speaking on the phone, you knew the call was over when you heard the dial tone after he had hung up. So with all the small talk, I knew something special was coming during our meeting in May 2012. I just didn't know what it was, good or bad.

Reid had stood by me through the embarrassing congressional hearings and internal conflicts at the NRC. Now, almost six months later, the commission was functioning better. There were still disagreements, but the failed coup had left my colleagues somewhat humbled, although no more willing to work collaboratively. The absence of an international crisis and the approval of the new reactor licenses also helped to ease the tension, although domestic problems with nuclear reactors were once again popping up. The most significant was the failure of a major piece of equipment at the San Onofre plant in southern California.

Ironically the troubled plant was in the district of Congressman Darrell Issa, who had held the House hearing the previous December, attacking my leadership of the agency. I invited him to join me that April on a tour of the facility, as I did for all representatives with a nuclear power plant in their district. Surprisingly he accepted. Although we spoke only briefly, our conversation was polite and professional. He acted as if he had never tried to publicly humiliate me.

Six weeks later, with the San Onofre plant problems behind me (although the plant would end up shutting down permanently), I was back in D.C., sitting across from Reid, who'd finally stopped chatting and come to the point.

"I wanted to have you announce that you intended to leave sometime before the end of your term next June," he told me. "Your successor would have been confirmed now along with Commissioner Svinicki but wouldn't start until next year, once you stepped down. But we spoke to the attorney general and he said the president can't appoint someone to start after the end of his term." President Obama would be facing reelection that fall. "We need you to act now."

Perplexed, I asked, "Are you asking me to resign?"

"Yes."

"How soon?"

"Monday."

Monday was just one workday away.

I knew Reid was asking me to resign because he saw an opportunity to prolong his influence over the commission, an opportunity he did not think he would have a year later, when my fixed five-year term would end and any pro-industry senator could simply block my replacement, forcing the president to pick a chairman from the people already on the commission. Acting now, Reid could use the industry's desire to confirm Kristine Svinicki to convince them to accept his replacement for me. He wanted to take this opportunity to ensure that someone he trusted would lead the agency for the next five years.

I admit I was disappointed. But I appreciated the opportunity Reid had given me to make a difference in government, and I was moved by his vigorous defense of me when my reputation took a hit. As I listened to him now, the thought of creating a new career started to sound more and more appealing. Leigh Ann and I had just found out she was pregnant with our first child. Being able to transition to a new life with the hope of more money, new and interesting work, and significantly less stress seemed like a good opportunity. Plus I knew Reid was telling, not asking. So I agreed.

The hardest part was telling my staff, who had supported and defended me during my time as chairman. It was an honor to have worked with them. As I gathered them around my conference table the next day to explain the situation, my lip quivered and I felt unwanted tears in my eyes.

I knew that there were many issues that would take years to resolve, if the commission had the will. The long-term Fukushima reforms needed more thought and to be put into action. New nuclear plants needed to be planned with foresight. Existing reactors needed to be monitored with vigilance and fortitude. And we would all have to think hard about how we wanted to create the energy that powers our world.

In one of my last speeches as chairman, I laid out my vision for the future of nuclear power, knowing that it was unlikely to end anytime soon. As I saw it, the industry could proceed along either of two paths. The first would lead to sustained growth and an increased appreciation for the value of nuclear energy, with a renewed focus on safety and a recognition of the real hazards of the technology. The second path would lead to a dead end; there would be decreased public and economic support for nuclear power due to an unwillingness to recommit to safety. As the years have passed, the industry missed an opportunity to embark on the first path and is pursuing that dead end.

My three and a half years as chairman of the NRC took place primarily during Obama's first term. His administration strove to restore some of the balance between the power of businesses, especially the large enterprises that dominate our lives as today's energy titans do, and that of the people. But even in this environment the nuclear industry held sway. Many in Congress, vulnerable to the industry's lobbying and to its ability to create jobs, were loath to oppose nuclear power. Soon the pressure to bolster the industry will become even more extreme because the economic case for nuclear power is slowly falling apart.

That's because nuclear power is one of the most expensive ways to generate electricity. Over the decades the cost of operating nuclear power plants has remained stable or even increased, unlike wind and solar and other sources of electricity, which have decreased dramatically in cost. Back when the Obama administration was championing the fight against climate change, nuclear power was touted despite its costs. Nuclear power does, after all, create little to no air pollutants; it does not spew particulate matter, mercury, sulfur dioxide, or any other typical by-product of fossil fuel combustion into the air. More important, nuclear plants emit no carbon dioxide or other greenhouse gases into the atmosphere. Still, it is becoming ever clearer that there are better, cheaper ways to create energy and combat climate change, especially as the marketplace continues to make renewable energy and energy efficiency more affordable.

Today the motivation for continuing to rely on this controversial form of electricity generation stems primarily from the powerful companies that have already made billions of dollars from this technology and the need for a source of power that does not emit significant amounts of greenhouse gases. But the question then becomes: What—considering the realities I've seen—will become of this technology in the future?

For the technology to survive, any future nuclear program must address three things. First and foremost, we must acknowledge that accidents will continue to happen. If more and more plants come online, more and more accidents are inevitable. To truly meet the needs of climate change abatement using nuclear power, the number of plants worldwide would have to rise from four hundred to a figure in the thousands. At those numbers, instead of happening once every ten years, a nuclear power plant accident could become an annual occurrence. Those incidents might not be of the same magnitude as Fukushima, but they would still create political and financial challenges. After all, the Three Mile Island accident did not release significant amounts of radiation, but it dramatically altered the nuclear power program in the United States.

It is easy to envision the United States, with its technological sophistication and political agility, coping successfully with an accident. But what about Vietnam or Niger? What about the countries that pursue nuclear power throughout the Middle East? Even Japan, one of the most advanced nations in the world, struggled in the aftermath of Fukushima. The prospect of nuclear power plant accidents will continue to challenge all nations.

Second, the cost of nuclear power continues to rise. Massive concrete domes or other structures are required to contain the radioactive material, and all that concrete must be paid for. Keeping a reactor cool requires much steel pipe, many motors, and a lot of pumps, each with its own price tag. Plants must be surrounded by large safety zones, land that must be bought and paid for. And still these expensive features cannot and likely never will prevent all accidents. Yet they must be part of designs for all nuclear power plants.

And so nuclear power remains one of most expensive sources of carbon-free electricity generation—maybe the most costly of all. The new reactors in Georgia and South Carolina, which I opposed during my time at the NRC, have failed to be constructed on time and within budget. The South Carolina plants have been canceled. And although they were supposed to be operating by 2016, the Georgia plants are years away from completion, and their cost has risen by billions of dollars. In order to save the flailing projects, Westinghouse, which designed the plants, bought out the companies primarily responsible for the construction in hopes of better controlling costs and schedule. Westinghouse proved no better than its predecessors. The massive delays and out-of-control expenses forced Westinghouse into bankruptcy. In the end the American taxpayer may be forced to subsidize this misadventure.

Existing plants are faring no better. The ever-decreasing cost of renewable energy and natural gas has placed great stress on the nuclear power facilities now in operation. Since I left the agency, more than ten reactors have shut down or plan to shut down because they are no longer financially viable. For the remaining reactors, as the pressure to cut costs increases, safety will suffer. The workforce will shrink, leaving fewer people available to identify problems. The maintenance of existing equipment will become more sporadic as activities are deferred to save money. The industry will likely promise these savings will come without any decline in safety, but the history of nuclear power shows this is not an easy commitment to keep. Such actions have been shown over the years to be the indicators of declining safety performance.

Third, a new approach to generating electricity with nuclear fission could appear. The physical process that drives a nuclear power plant—the fissioning of uranium atoms—is an extremely efficient means of creating energy. But harnessing that energy creates problems that new technology might solve. The decision decades ago to use water to control the heat of fission and transfer this heat to a turbine via steam has saddled nuclear power plants with a problem: nearly all the flaws in the safety systems needed to control a plant after an accident stem from the inherent weaknesses of water. If you replace the water with gases or liquid metals, you could build a safer reactor—in theory, anyway. Unfortunately the resources to bring such a plant to commercial production are very costly and the time needed to get there very long. And these new plants would surely come with their own safety issues. It is hard for me to believe that in forty years, when a new generation of non-water-based nuclear reactors may become viable, we will not have harnessed our thirst for electricity in a far more efficient manner.

Nuclear power is a model of electricity production that dates from the early part of the twentieth century. It is a system characterized by large power plants that produce large amounts of power and transmit it over large distances. The system arose during a period in which plants—coal- and nuclear-powered ones especially—were built by state-sanctioned monopolies with captive financiers, the people who paid for the electricity from their state-regulated utility. But those days are gone. Advances in technology and manufacturing have made other types of clean electricity feasible and much less costly. In addition, many states have turned to open electricity markets in which the state-sanctioned utility with its monopolies must compete against newer, cheaper alternatives. They do not fare well. The nuclear industry, with its monolithic plants over budget and behind schedule, does not fit the needs of the twenty-first century.

One solution to this problem is to go local, which is how much renewable electricity works best. Instead of screwing solar panels on top of existing roofs, homes will eventually be built with solar roofing material, as well as solar siding. We will likely begin to think of electricity much as we do hot water: as something we make in our homes on demand. Can a house of the future use solar hot water, as well as solar roof and building materials, to generate a portion of its electricity? I think that's very possible. We will still likely need some form of power plant to generate a baseline supply of power for more energy-intensive facilities like hospitals and factories, but the need to transmit large amounts of power from such plants will diminish. There will also be a place for larger renewable facilities, and I believe that in forty years we will have figured out how to fix the problems these pose today. New technology will make the grid more adaptable and agile.

Nuclear power does not work well in this new model of distributed, agile generators. It is large and bulky and will lumber into extinction. Yet no matter how much new technology might renew our energy system, nuclear power will remain with us for some time. Despite the recent shutdowns of several nuclear power plants and the failure to complete those that the NRC licensed while I was chairman, dozens of nuclear power plants will continue to operate in America for the next fifteen years at least. In the decades ahead we will need to focus on keeping these plants safe. Probability cannot deceive us into believing that the hazards of nuclear power are small or insignificant. Breaking through the political support for nuclear power and the industry's defenses will be difficult and costly, but it is necessary to ensure that the autumn years of nuclear power pass with as few incidents as possible. There are no guarantees that a plant will be safe, nor are there guarantees a plant will be dangerous. But the past has shown that the hazards are real.

Almost a decade after leaving the agency, I am still haunted by two images. The first is all that water flowing downstream from a massive snowmelt, flooding the Missouri, fanning out sideways to turn the surrounding farmland on both sides of the river into an inland sea, and squeezing the nuclear plant at Fort Calhoun. The second image is the idle plants at Fukushima, including the one whose reactors are now a ruin of steel rods, the radioactive poison buried deep inside the site still beyond the control of the engineers. I started my life as a scientist in awe of humans' ability to see the genius of nature and harness it. I left my job as a nuclear regulator humbled at what nature can do to turn our technological inventions against us.

APPENDIX

A Peek at the Science of Nuclear Reactors

The starting point for understanding nuclear power is through its product: electricity. We notice it mostly when the power fails or our cell phone loses its charge; nothing reminds us more of our reliance on electricity than sitting on a dirty airport carpet huddled by a power outlet. But electricity—which humans have harnessed for only 150 years—is one of nature's phenomena that could easily be confused with magic. Lightning crashes from the clouds, turning a dark night into bright day. A compass needle turns toward the North as you walk in a circle. U-shaped magnets clamp together in a fierce hold.

Given our ever-escalating, near-addictive need for electricity and the greenhouse gases that most sources of electricity emit, producing electricity with limited amounts of pollution has become one of the most pressing technological needs of our time. Nuclear power appeared to provide the ideal solution.

Most electricity-generating plants operate with the same basic approach. They heat water into steam and slam the steam into massive steel blades. The force of the steam turns the blades just like inside a jet engine. Attached to the shaft that spins inside these turbine blades is a large magnet. The wonders of electricity produce an electric current in the wires surrounding that magnet. From there, the electricity is transmitted to homes and businesses throughout the country.

The reactor is hidden deep within the plant, inside protective buildings and behind rows of high-tech security fences. If you could see past all these barriers straight into the reactor core, you would see a giant metal cylinder filled with water, with thousands of pencil-thin rods suspended inside. These rods (called fuel pins) are composed of small cylindrical pellets of a ceramic form of uranium, a heavy metal found naturally in the earth.

Much the way a Revolutionary War soldier would stuff gunpowder into a musket, these pellets—each about the size of the end of your thumb—are pressed firmly into the fuel pins, the uranium fitting snugly into a protective metal sheath. Each fuel assembly has hundreds of these fuel pins, each about twenty feet long, arranged in square lattices, with every nuclear reactor featuring hundreds of such fuel assemblies. Through more physics magic, these thin rods of uranium create the heat that is needed to make electricity.

To understand the source of atomic power, you have to travel deeper inside these already small fuel pellets, into the uranium atoms they comprise. An atom of uranium is so small that the ratio of the size of a bicycle to the size of its nucleus would be about the same as the ratio of the distance to the nearest star to the distance between the bicycle's wheels. This is a very tiny space in which to operate. But nature uses a very specific process in these tiny spaces: nuclear fission.

Nuclear fission is the splitting of a uranium atom—a relatively heavy atom—into two lighter atoms. Like sparks from a live wire, nuclear fission also shoots out even smaller particles, aptly named subatomic particles. The two most interesting are neutrons and photons. Neutrons are a type of particle found in the nucleus of atoms that play a very special role in sustaining the nuclear fission process.

Inside a power plant, nuclear fission starts with neutrons from some outside source bombarding the uranium atoms in the fuel pins. This triggers the rupture described above, with uranium atoms splitting into two lighter atoms and some subatomic particles. The new neutrons released from these uranium atoms can then start the process again. Like a No. 1 bowling pin knocked over for a perfect strike, the neutrons start a cascade of fissioning uranium atoms that can continue in the reactor core for years.

The other particles created by nuclear fission—the photons and smaller atoms—are the seeds of electricity in a nuclear power plant. (Photons are just another way to talk about light, albeit often light that is very different from what our eyes can see.) These atoms and photons possess a tremendous amount of energy for such small particles, enough to heat up the surrounding material ever so slightly. When aggregated in a typical nuclear reactor, these small bursts of energy can produce enough electricity to power thousands of homes.

The smaller atoms produced from each uranium atom through nuclear fission are also radioactive. It is this radioactivity that is responsible for many of the safety problems at nuclear power plants. If left unchecked, these hazardous nuggets have sufficient energy to melt and blast their way out of the reactor vessels.

To imagine how this whole process works, think of the

reactor core as rows and rows of school buses in a parking lot. Each of these rows is a fuel pin, and each school bus is a uranium atom. As nuclear fission happens, some of the buses change into two other vehicles. One bus might become a sport utility vehicle and a bicycle. Another might become two small Volkswagens. These new vehicles would still be squeezed into the parking spot of the school bus that created them. But now there might be a little more space for them to move. This space would be the extra energy released by the atomic split. Surrounding the parking lot is a secure fence that keeps the vehicles inside. In a nuclear accident, the parking lot fence would come down and the vehicles would speed off in all directions, some carrying dangerous cargo (radiation).

The recipe for disaster is simple: if the fuel overheats, an accident happens. During normal operations, the primary source of heat comes from nuclear fission. Because fission is dependent on neutrons, the heat from nuclear fission is very easy to stop: you simply add materials to the reactor to absorb the neutrons. Plants in the United States have systems in place to ensure that runaway nuclear fission cannot occur.

A second major source of heat comes from the natural radioactive decay of the atoms created by nuclear fission. This energy can last for a long time after nuclear fission has stopped. There is no way to curtail this natural radioactive decay. Think of the electric burners on a stove; even though the knob may be turned to the off position, the burners stay hot enough to burn you for some time. During normal operations, removing this decay heat takes hours. During an accident in which all safety systems fail, this heat would destroy the reactor in a matter of hours, releasing hazardous radiation to distances up to tens of miles from the plant.

To manage the heat, a massive system of pipes and pumps moves water around the plant, either taking heat away from the reactor fuel and releasing it into the environment, or converting the heat into steam and electricity. When the cooling systems function the way they are supposed to, heat is removed tiny bit by tiny bit. Imagine being a molecule of water riding through the reactor core picking up little bits of heat. The first bundle of heat you collect leaps off the reactor fuel. Better than the most extreme amusement park ride, you blast upward dozens of feet, carrying your packet of heat safely away from the reactor. At the top of the steel vessel that contains the reactor fuel, you're forced around a sharp turn and into one of the plant's large main pipes. Bending and twisting through a series of turns much like an extreme water slide, you get a burst of energy from a gigantic pump that keeps the water flowing through the system. This burst of energy propels you to your first stop, a massive structure with thousands of tiny tubes; these are the steam generators. On one side of these tubes is clean, nonradioactive water. You send off your heat packet through a thin metal tube into a separate system of clean water. This second water system ultimately deposits the heat into a river or lake.

This finely tuned, powerful process is repeated in hundreds of plants around the world with very little trouble. Sometimes, however, something goes wrong. The pumps that circulate the water may fail. The supply of power to the motors that turn the pumps may drop. The pipes containing radioactive water may leak. If backup safety systems fail to flood the reactor in time, the heat will eventually over-heat the fuel, until it gets hot enough to melt the dangerous radioactive materials and allow radiation to flee into the environment.

But nuclear plant designers were aware that this could happen, so they designed systems to trap the radioactive materials in the reactor. The first barrier is the casing surrounding the uranium fuel pellets. Under normal conditions, this thin metal wall keeps the radioactive materials contained. But this barrier can quickly fail. The next line of defense is the network of thick pipes that contain the water used to cool the reactor. This system is a closed network, keeping the radioactivity contained. But this system too can fail. Pipes can leak (as can happen during an accident due to intense pressure from the incredibly hot steam) or break under the stress of operating in such a harsh environment—or during an earthquake.

If all these systems fail, the final protection comes from a structure known as the containment. In most plants in operation, the containment is a massive concrete dome encasing the safety and energy-generating systems. Some of these structures are nearly large enough to hold the dome of the U.S. Capitol building. They can be over a hundred feet high and a hundred feet in diameter. The role of this final line of defense is to absorb the pressure produced by the steam generated by an out-of-control reactor. Think of it as a giant teapot built to hold boiling water for a very long time. But in some accident situations, even this structure will fail.

In a devious twist of nature, when the steam inside the reactor core interacts with the metal sheath of the fuel pins, the resulting chemical reaction produces volatile and explosive hydrogen, which lacks only a source of oxygen to ignite. In the presence of oxygen, hydrogen gas will explode with a force capable of destroying many of the containment's thick concrete walls. As the United States slept, many in the rest of the world watched as the first of three hydrogen explosions at the Fukushima reactor site hurled obliterated walls high into the sky. But unlike the *Hindenburg* crash that ended the pursuit of hydrogen power, this did not seal the fate of nuclear technology.

The real danger of a nuclear power plant is not the destruction of the plant itself, but rather the harmful effects of the radiation emitted by the materials released. For a long time after the discovery of radiation, these health hazards remained hidden. Marie Curie, the Nobel Prize winner who discovered much of what we now describe as radiation, died from a rare disease likely the result of repeated radiation exposure. It would take decades and a devastating world war for us to truly begin to understand the impacts of exposure to radiation.

From studies of Japanese survivors of Hiroshima and Nagasaki, scientists established thresholds of radiation exposure considered safe, as well as harmful levels of exposure that would incite acute fatal illnesses. In between these two extremes the evidence generally shows that repeated exposure to radiation increases the chance that people will develop diseases, primarily cancer.

Radiation is typically carried by three types of energy: alpha, beta, and gamma rays or particles. (Sometimes these forms of energy are referred to as particles, sometimes as rays or waves; the distinction is important only to physicists and some philosophers.) Gamma rays are just a type of light, one that has a lot more energy than the type we're used to seeing. Beta particles are types of electrons, and alpha particles are pieces of helium atoms. Each of these types of radiation damages the human body in different ways. Gamma radiation often penetrates deeply, degrading health even when a person is far from the radiation's source. Alpha particles cannot pass through the skin; they have to be ingested to do any harm.

When these particles come in contact with the body, they can damage cells and tissues, making people sick or susceptible to diseases like cancer. If the exposure is high enough, the damage can quickly lead to death. If the exposure is smaller and sustained over a long period of time, cells in the body may become damaged and grow into cancer cells. Physics, biology, and medicine have not yet determined exactly how much radiation triggers cellular changes that will lead to cancer. But we know that it happens and that above a certain level of exposure, a person will die.

How does radiation come into contact with people? In general, radioactive elements behave just like nonradioactive ones. When nature is assembling the building blocks of our foodstuff, our water, and our air, it picks whatever elements are on hand. After the accident at Fukushima, there were a lot more radioactive elements around than usual. Some were common elements that can become the building blocks of basic foods and lead to radiation exposure once ingested. Some, like iodine, can be absorbed into your thyroid; cesium, another dangerous element, can become a part of your bones and emit radiation from there. Radioactive material can settle on the ground around your house or your children's school, exposing them to penetrating gamma radiation from a distance. Sometimes only thick lead walls can stop radiation from penetrating the body. So preventing the spread of radioactive contamination on land is essential.

NOTES

Chapter 2: Forget and Repeat: A Brief but Necessary History of Accidents

26 *It started on March 28*: U.S. Nuclear Regulatory Commission, Office of Public Affairs, *Backgrounder: Three Mile Island Accident* (Washington, DC, 2013), 1. Specific details of the accident are pulled from this document and the Kemeny Report narrative.

27 *Seeing the pressurizer appear to go solid*: Ibid., 2.

27 *Outside the walls*: Ibid.

28 *"I don't believe in my mind"*: John G. Kemeny et al., *Report of the President's Commission on the Accident at Three Mile Island* (Washington, DC, 1979), 103.

29 *a small hydrogen explosion occurred*: Ibid., 107.

30 *Nonetheless nuclear analysts*: Ibid., 114.

33 *on the soles of a worker's shoes*: Malcolm W. Browne, "Swedes Solve a Radioactive Puzzle," *New York Times*, May 13, 1986.

33 *Chernobyl got little attention*: U.S. Nuclear Regulatory Commission, *Report on the Accident at the Chernobyl Nuclear Power Station* (Washington, DC, 1987), 1.

34 *Thirty emergency workers died*: United Nations, Scientific Committee on the Effects of Atomic Radiation, *Sources and Effects of Ionizing Radiation: Volume II, Annex D* (New York, 2011), 58.

34 *But it is likely that*: Ibid., 64.

34 *Five million people living*: Ibid., 53.

34 *these cancers could have been prevented*: Ibid., 66.

35 *One of these metals, Alloy 600*: Ibid., 10.

36 *NRC's initial assessment*: U.S. Nuclear Regulatory Commission, *Davis-Besse Reactor Vessel Head Degradation Lessons-Learned Task Force Report* (Washington, DC, 2002), 2.

Notes

Chapter 3: The Burning Issue: The Battle to Prevent Nuclear Fires

45 *On one side were the traditional nuclear safety standards*: The terms "deterministic," "risk-informed," and "performance-based" were created by the U.S. Nuclear Regulatory Commission to classify the various old and new approaches to dealing with the hazards from nuclear power plants. The commission formally defined them in 1999. See U.S. Nuclear Regulatory Commission, Staff Requirements - Secy-98-144 - White Paper on Risk-Informed and Performance-Based Regulation (Washington, DC, 1999).

52 *"NRC staff resources inefficiently"*: Alexander Marion to Eric Leeds, November 15, 2010, U.S. Nuclear Regulatory Commission Public Library.

Chapter 5: Accidents Do Happen: The Tragedy of Fukushima

71 *"nuclear village"*: This term was used by several of the reports done by Japanese organizations following the accident. See Mindy Kay Bricker, *The Fukushima Daiichi Nuclear Power Station Disaster: Investigating the Myth and Reality* (London, 2014), and *The Official Report of the Fukushima Nuclear Accident Independent Investigation Commission: Executive Summary* (Tokyo: National Diet of Japan, 2012).

Chapter 7: Tsunamis in the Heartland: A Scenario for an American Fukushima

106 *"The design bases for these structures"*: Many of the General Design Criteria are written without jargon, unlike most of the NRC's regulations. See Appendix A to Part 50—General Design Criteria for Nuclear Power Plants, *Code of Federal Regulations*, Title 10 (2018), 999–1007.

Notes

Chapter 8: Fukushima Effects: The Fight over Essential Industry Reforms

121   *the report . . . sent shock waves through the industry*: The Task Force report is very readable and provides a nice summary of the accident. See Charles Miller et al., *Recommendations for Enhancing Reactor Safety in the 21st Century: The Near-Term Task Force Review of Insights from the Fukushima Dai-Ichi Accident* (Washington, DC, 2011).

121   *"The current regulatory approach"*: Ibid., vii.

INDEX

Index

Index

nuclear power industry (*cont.*)

reforms mandated after Three Mile Island, 31–32

reforms recommended after Fukushima, 116–31

regulations and, 49

resistance to dealing with fire dangers, 47–53

as responsible for dealing with accidents, 76–77

safety resting with plant owners and operators, 110–11

spent fuel pools and, 91

U.S. technology and, 133

voluntary performance-based standards and, 49–50, 122, 128

See also nuclear power accidents

nuclear power plants

Alloy 600 problems, 35–37

canceling of new, following Three Mile Island accident, 32

community safety and, 135

cost of building, 135

declining safety performance, 165

diesel generators for, 78, 114

early mistakes in building, 35–36

hearing on new Georgia reactors, 138, 139

impact of Fukushima on new applications, 137, 138

jobs created by, 135

license condition, 154, 155–56

licensing of new plants, 69, 132–43

loan for new Georgia project, 136

plant designers, 105–6, 135, 136

plant locations, 134–35

public interest groups' opposition to, 137–38

safety protocol at, 106

safety resting with plant owners and operators, 110–11

shutdowns of, 165

sites for plants, 106, 111

See also specific plants

nuclear reactors, 7

Alloy 600 and, 35–37

argument for, 163

boric acid corrosion and, 37

catastrophic releases of radiation and, 116

"decay heat," 26

designers of, 135

earthquake risks, 106–7, 174

failure of fail-safe setups, 20–21, 39–40, 41, 174

fire danger, 41–53

flooding risk, 109–15

Index

ABOUT THE AUTHOR

Dr. Gregory Jaczko served as chairman of the U.S. Nuclear Regulatory Commission from 2009 to 2012, and as a commissioner from 2005 to 2009. As chairman, he played a lead role in the American government's response to the nuclear accident in Fukushima, Japan, in 2011. Jaczko is now an adjunct professor at Princeton University and Georgetown University, and an entrepreneur with a clean energy development company.